AP 微积分 AB/BC 指导用书
AP Calculus Study Guide

潘胤杰　张　梦　编

Billson International Ltd.

Published by
Billson International Ltd
27 Old Gloucester Street
London
WC1N 3AX
Tel:(852)95619525

Website:www.billson.cn
E-mail address:cs@billson.cn

First published 2025

Produced by Billson International Ltd
CDPF/01

ISBN 978-1-80377-150-2

AP Calculus Unit Guide & Lesson Handout

Landwave NA Group

July 25, 2024

澜大教育集团
LANDWAVE GROUP

澜大如同大树一样生根在国际教育领域的各个方面，以专业的教学系统和尖端团队，为北美留学、英联邦留学、国际学校择校备考、多语种教育培训、大学生留学规划等提供整体学习方案和专业的国际课程辅导，并与国际学校始终保持着密切的合作。今天的澜大，宛如承载着众多家庭期望的巨舰，在国际教育领域中——领引众志，行稳远航。

愿景	使命	价值观
做最具口碑影响力的 教育品牌	集教育梦想为怀，共建平台 以造就群英为愿，点亮未来	开放包容　坚韧乐观 谦逊感恩　精进自省 拥抱挑战　高效务实

Contents

1   Limit and Continuity

1.1   Introduction to Limits

Limit is a basic tool for studying **calculus**, which gives a a brand-new overview of the function by taking use of the concept *"approaching but not reaching"*.

Theorem 1.1 (The Definition of Limit).

If $f(x)$ becomes arbitrarily close to a value L as x approaches c from **either side**, then the limit of $f(x)$ is L.

$$\lim_{x \to c} f(x) = L.$$

If f is a **continuous function** on an open interval containing the number a, then

$$\lim_{x \to a} f(x) = f(a)$$

Consider the following function:

x	0.75	0.9	0.99	0.999	1	1.001	1.01	1.1	1.25
$f(x)$	2.313	2.710	2.970	2.997	?	3.003	3.030	3.310	3.813

$f(x)$ approaches 3 when x is approaching 1 from the left hand side.
$f(x)$ approaches 3 when x is approaching 1 from the right hand side, thus:

$$\lim_{x \to 1} f(x) = 3$$

This is read as "the limit of $f(x)$ as x approaches 1 is 3. "

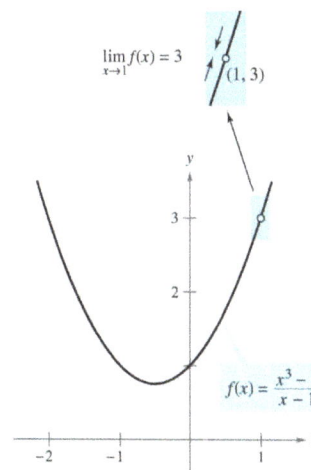

In the previous example, we notice that the *approaching value* from both sides needs to be taken into consideration.

Corollary 1.2 (One-Sided Limit).

If $f(x)$ approaches L as x approaches c from the right, then: $\lim_{x \to c^+} f(x) = L$

If $f(x)$ approaches L as x approaches c from the left, then: $\lim_{x \to c^-} f(x) = L$

Then we can judge the existence of a limit by checking whether the *one-sided limits* are the same.

Claim 1.3 (Existence of a Limit).

If $\lim_{x \to c^+} f(x) \neq \lim_{x \to c^-} f(x)$, then $\lim_{x \to c} f(x)$ does not exist (DNE).

If $\lim_{x \to c^+} f(x) = \lim_{x \to c^-} f(x) = L$, then $\lim_{x \to c} f(x) = L$.

Example (1). For the following functions, find the limit of $f(x)$ at $x = a$.

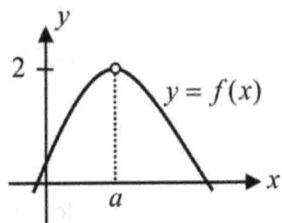

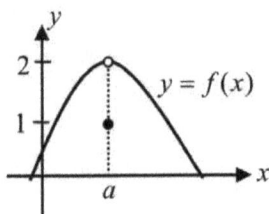

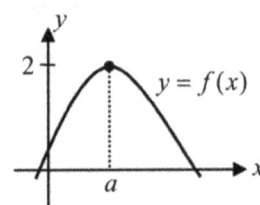

Solution.

Example (2). Determine $\lim\limits_{x \to 0} \dfrac{|x|}{x}$.

Solution.

Example (3). If $y = \begin{cases} e^{2x}, & -4 \le x < 0 \\ xe^x, & 0 \le x \le 4 \end{cases}$, find $\lim\limits_{x \to 0} f(x)$.

Solution.

1.2   Evaluating Limits

To evaluate the limits of functions and expressions, we need to clarify the properties of limits.

Theorem 1.4 (Properties of Limits).

Given $\lim\limits_{x \to a} f(x) = L$ and $\lim\limits_{x \to a} g(x) = M$ and L, M, a and c are real constants, then:

1. $\lim\limits_{x \to a} c = c$

2. $\lim\limits_{x \to a} c\,[f(x)] = c \lim\limits_{x \to a} f(x) = c \cdot L$

3. $\lim\limits_{x \to a} [f(x) \pm g(x)] = \lim\limits_{x \to a} f(x) \pm \lim\limits_{x \to a} g(x) = L \pm M$

4. $\lim\limits_{x \to a} [f(x) \cdot g(x)] = \lim\limits_{x \to a} f(x) \cdot \lim\limits_{x \to a} g(x) = L \cdot M$

5. $\lim\limits_{x \to a} \dfrac{f(x)}{g(x)} = \dfrac{\lim\limits_{x \to a} f(x)}{\lim\limits_{x \to a} g(x)} = \dfrac{L}{M}$, where $M \neq 0$

6. $\lim\limits_{x \to a} [f(x)]^n = \left(\lim\limits_{x \to a} f(x) \right)^n = L^n$

Common techniques in evaluating limits are:

(1) Substituting Directly

Example (1). Evaluate $\lim\limits_{x \to 2} \left(5x^2 - 3x + 1 \right)$

Solution.

Example (2). Evaluate $\lim\limits_{x \to \pi} 3x \cdot \sin x$

Solution.

Example (3). Evaluate $\lim\limits_{x \to 1} \dfrac{3x^2 + 2x - 1}{x^2 + 1}$

Solution.

(2) Factoring and Simplifying

> **Example (4).** Evaluate $\lim\limits_{t\to 2} \dfrac{t^2 - 3t + 2}{t - 2}$
>
> Solution.

> **Example (5).** Evaluate $\lim\limits_{x\to b} \dfrac{x^5 - b^5}{x^{10} - b^{10}}$
>
> Solution.

> **Example (6).** Evaluate $\lim\limits_{x\to -2} \dfrac{x^3 + 8}{x^2 - 4}$
>
> Solution.

(3) Conjugate Radicals

> **Example (7).** Evaluate $\lim\limits_{t\to 0} \dfrac{\sqrt{t + 2} - \sqrt{2}}{t}$
>
> Solution.

> **Example (8).** Evaluate $\lim\limits_{x\to 1} \dfrac{x - 1}{\sqrt{x + 3} - 2}$
>
> Solution.

(4) Using Squeeze Theorem

> **Theorem 1.5** (Squeeze Theorem).
>
> If f, g, and h are functions defined on some open interval containing a such that $g(x) \leq f(x) \leq h(x)$ for all x within the interval except possibly at a itself, then:
>
> $$\text{if } \lim_{x \to a} g(x) = \lim_{x \to a} h(x) = L, \quad \text{then } \lim_{x \to a} f(x) = L$$

Example (9). Evaluate $\lim_{x \to 0} \left(x^2 \, \sin \left(\dfrac{\pi}{x} \right) \right)$

Solution.

(5) Other Basic Limits

> **Lemma 1.6.** If x is measured in radians, $\lim_{x \to 0} \dfrac{\sin x}{x} = 1$, and $\lim_{x \to \infty} x \cdot \sin \left(\dfrac{1}{x} \right) = 1$

Example (10). Evaluate $\lim_{x \to 0} \dfrac{\sin 3x}{x}$

Solution.

Example (11). Evaluate $\lim_{y \to 0} \dfrac{y^2}{1 - \cos y}$

Solution.

(6) Limits involving composite functions

Theorem 1.7.

If f and g are functions such that $\lim\limits_{x \to c} g(x) = L$ and $\lim\limits_{x \to L} f(x) = f(L)$, then:

$$\lim_{x \to c} f(g(x)) = f\left(\lim_{x \to c} g(x)\right) = f(L)$$

The previous theorem states the simplest case that $f(x)$ has **identical one-sided limit** at $x = L$. However, if we have $\lim\limits_{x \to L^-} f(x) \neq \lim\limits_{x \to L^+} f(x)$, then we need to check the *direction of approach*.

Example (12). Evaluate the following limit: $\lim\limits_{x \to 2} f(g(x))$, by the following graph.

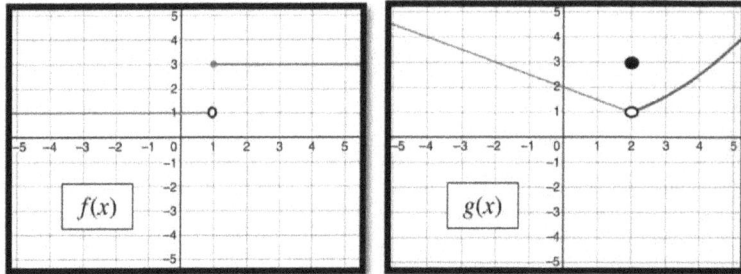

Solution.

Example (12).

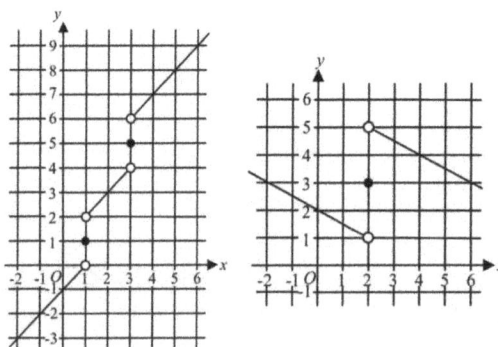

Graph of f   Graph of g

The graphs of the functions f and g are shown. What is $\lim\limits_{x \to 2^-} f(g(x))$?

Solution.

1.3   Limits Involving Infinities and Asymptotes

Theorem 1.8 (Infinite Limits as $x \to a$).

If f is a function defined at every number in some open interval containing a, except possibly at a itself, then:

(1) $\lim\limits_{x \to a} f(x) = +\infty$ means that $f(x)$ increases *without bound* as x approaches a.

(2) $\lim\limits_{x \to a} f(x) = -\infty$ means that $f(x)$ decreases *without bound* as x approaches a.

The simplest example for this limit is $f(x) = \dfrac{1}{x}$, we may find that:

$$\lim_{x \to 0^+} \left(\frac{1}{x} \right) = \infty, \qquad \lim_{x \to 0^-} \left(\frac{1}{x} \right) = -\infty$$

Then we can discuss about the limit of $\dfrac{f(x)}{g(x)}$ with $g(x)$ approaches 0.

Corollary 1.9 (Limits with Zero Denominator).

If $\lim\limits_{x \to a} f(x) = c$, where $c > 0$, and $\lim\limits_{x \to a} g(x) = 0$, then:

$$\lim_{x \to a} \frac{f(x)}{g(x)} = \begin{cases} \infty & \text{if } g(x) \text{ approaches 0 through positive values} \\ -\infty & \text{if } g(x) \text{ approaches 0 through negative values} \end{cases}$$

If $\lim\limits_{x \to a} f(x) = c$, where $c < 0$, and $\lim\limits_{x \to a} g(x) = 0$, then:

$$\lim_{x \to a} \frac{f(x)}{g(x)} = \begin{cases} -\infty & \text{if } g(x) \text{ approaches 0 through positive values} \\ \infty & \text{if } g(x) \text{ approaches 0 through negative values} \end{cases}$$

Example (1). Evaluate the limits: (a) $\lim\limits_{x \to 2^+} \dfrac{3x - 1}{x - 2}$ and (b) $\lim\limits_{x \to 2^-} \dfrac{3x - 1}{x - 2}$

Solution.

Example (2). Evaluate $\lim\limits_{x \to 5^-} \dfrac{\sqrt{25 - x^2}}{x - 5}$

Solution.

Example (3). Evaluate $\lim\limits_{x \to 2^-} \dfrac{\lfloor x \rfloor - x}{2 - x}$, where $\lfloor x \rfloor$ is the greatest integer value of x.

Solution.

Theorem 1.10 (Limits at infinity as $x \to \pm\infty$).

If f is a function defined at every number in some interval (a, ∞), then $\lim\limits_{x \to \infty} f(x) = L$ means that L is the limit of $f(x)$ as x *increases without bound*.

If f is a function defined at every number in some interval $(-\infty, a)$, then $\lim\limits_{x \to -\infty} f(x) = L$ means that L is the limit of $f(x)$ as x *decreases without bound*.

From the previous theorem, we may deduce the limit of a rational function $f(x) = \dfrac{p(x)}{q(x)}$.

Corollary 1.11 (Limit at $x \to \infty$ of Rational Functions).

Let $\delta(p)$ and $\delta(q)$ denote the degree of the numerator $p(x)$ and denominator $q(x)$ respectively for:

$$f(x) = \frac{p(x)}{q(x)}$$

1) If $\delta(p) < \delta(q)$, then $\lim\limits_{x \to \infty} \dfrac{p(x)}{q(x)} = 0$.

2) If $\delta(p) > \delta(q)$, then $\lim\limits_{x \to \infty} \dfrac{p(x)}{q(x)} = \infty$ or $-\infty$.

3) If $\delta(p) = \delta(q)$, then $\lim\limits_{x \to \infty} \dfrac{p(x)}{q(x)} = c$, where c is the ratio of the leading coefficients of $p(x), q(x)$.

Example (4). Evaluate $\lim\limits_{x \to \infty} \dfrac{6x - 13}{2x + 5}$.

Solution.

Example (5). Evaluate $\lim\limits_{x \to \infty} \dfrac{2x + 1}{\sqrt{x^2 + 3}}$.

Solution.

Example (6). Evaluate $\displaystyle\lim_{x \to \infty} \frac{1 - x^2}{10x + 7}$.

Solution.

The limit at $x \to \pm\infty$ can also help us find the asymptotes of a function.

Theorem 1.12 (Horizontal and Vertical Asymptotes).

A line $x = b$ is called a horizontal asymptote for the graph of a function f if either

$$\lim_{x \to \infty} f(x) = b \qquad \text{or} \qquad \lim_{x \to -\infty} f(x) = b.$$

A line $x = a$ is called a vertical asymptote for the graph of a function f if either

$$\lim_{x \to a^+} f(x) = \pm\infty \qquad \text{or} \qquad \lim_{x \to a^-} f(x) = \pm\infty.$$

Note: the one-sided limit are not necessarily to be identical.

Example (7). Find the horizontal and vertical asymptotes of the function $f(x) = \dfrac{3x + 5}{x - 2}$.

Solution.

Example (8). Find the horizontal and vertical asymptotes of the function $f(x) = \dfrac{2e^x - 1}{3 - 5e^x}$.

Solution.

Example (9). Using your calculator, find the horizontal asymptotes of the function $f(x) = \dfrac{2\sin x}{x}$.

Solution.

1.4 Continuity of a Function

In some ways, we can use the limit to determine the continuity of a function at certain point.

> **Theorem 1.13** (Continuity of $f(x)$ at a Point).
>
> A function $f(x)$ is continuous at c, if:
>
> 1. The limit of $f(x)$ exists at $x = c$: $\lim\limits_{x \to c} f(x) = L$
>
> $$\lim_{x \to c^-} f(x) = \lim_{x \to c^+} f(x) = L$$
>
> 2. The function is defined at $x = c$: $f(c)$ exists
>
> 3. $\lim\limits_{x \to c} f(x) = f(c) = L$

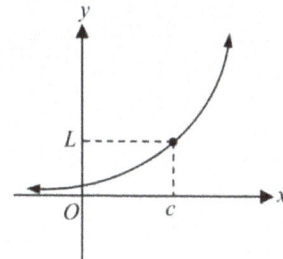

We can say $f(x)$ is continuous at $x = c$ when all three properties holds. If any one of them is not satisfied, then the function is considered to be **discontinuous** at $x = c$. The type of **discontinuity** is based on which property is not satisfied.

> **Corollary 1.14** (Removable and Non-Removable Discontinuity).
>
> **Removable Discontinuity**
>
> A function has **removable discontinuity** at $x = c$, if f can be **made continuous** by filling in a point (or changing the value of $f(c)$). The point of discontinuity is a removable discontinuity at $x = c$.

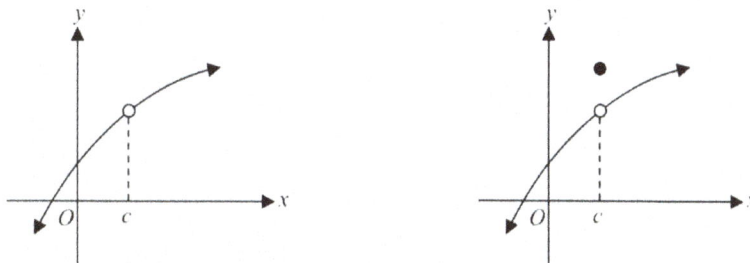

A function f has a **removable discontinuity** at $x = c$ if $\lim\limits_{x \to c} f(x)$ exists but:

$$f(c) \text{ does not exist} \qquad \textbf{OR} \qquad f(c) \neq \lim_{x \to c} f(x)$$

Non-removable Discontinuity

A function has **non-removable discontinuity** at $x = c$, if there is NO WAY to make f to be continuous by filling a point (or changing the value of $f(c)$).

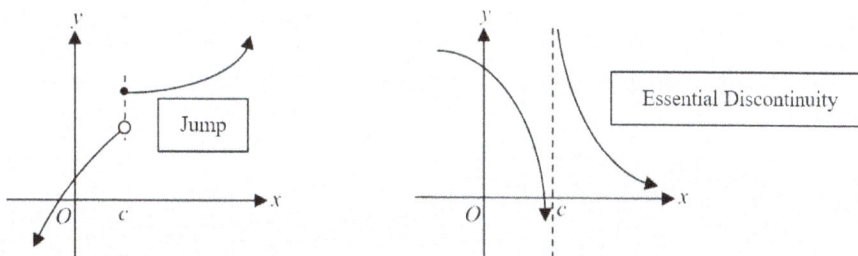

A function f has a **jump discontinuity** if $\lim\limits_{x \to c} f(x)$ does not exist ($\lim\limits_{x \to c^+} f(x) \neq \lim\limits_{x \to c^-} f(x)$).

A function f has an **essential discontinuity** if either $\lim\limits_{x \to c^-} f(x) = \pm\infty$ or $\lim\limits_{x \to c^+} f(x) = \pm\infty$. In this case, the graph of function f may have a vertical asymptote.

11

Example (1). Find the points of discontinuity of the function $f(x) = \dfrac{x+5}{x^2 - x - 2}$

Solution.

Example (2). Determine the intervals on which the given function is continuous:

$$f(x) = \begin{cases} \dfrac{x^2 + 3x - 10}{x - 2}, & x \neq 2 \\ 10, & x = 2 \end{cases}$$

Solution.

Example (3). For what value of k is the function $f(x) = \begin{cases} \dfrac{5\sin x}{x}, & x < 0 \\ k - 3x, & x \geq 0 \end{cases}$ continuous at $x = 0$?

Solution.

Example (4). For what value of k is the function $g(x) = \begin{cases} \dfrac{x^2 - k^2}{x - k}, & x \neq k \\ 10, & x = k \end{cases}$ continuous at $x = k$?

Solution.

Theorem 1.15 (Intermediate Value Theorem).

If f is continuous on a closed interval $[a, b]$ and k is any number between $f(a)$ and $f(b)$, then there exists **at least one** number of c within the interval (a, b) such that $f(c) = k$.

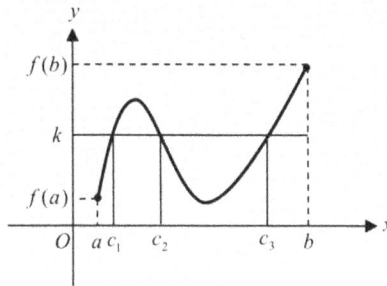

Example (5). A function f is continuous on $[0, 5]$, and some of the values of f are shown below.

x	0	3	5
f	-4	b	-4

If $f(x) = -2$ has no solution on $[0, 5]$, then b could be:

(A) 1   (B) 0   (C) -2 (D) -5

Solution.

Example (6). The function f is continuous on the closed interval $[-5, 5]$ and have values that are given in the table below. What is the minimum number of times that the function takes on the value $f(x) = -2$ on the interval $[-5, 5]$?

x	-5	-3	-1	1	3	5
$f(x)$	4	2	-3	1	-4	-5

(A) 1   (B) 2   (C) 3   (D) 4

Solution.

2 Differentiation and Fundamental Properties

2.1 Definition of Derivatives

The derivative of a function is one of the basic concepts of mathematics. The process of finding the derivative is called differentiation. The derivative of a function can also be considered as *instantaneous rate of change.*

Theorem 2.1 (Average and Instantaneous Rate of Change).

The **average rate of change** of y (slope m) with respect to x over the interval $[x_1, x_2]$ or $[x_1, x_1 + h]$ is given by

$$r_{avg} = \frac{\Delta y}{\Delta x} = \frac{y_2 - y_1}{x_2 - x_1} = \frac{f(x_2) - f(x_1)}{x_2 - x_1} = \frac{f(x_1 + h) - f(x_1)}{h}$$

It also represents the slope of the **secant line** ℓ.

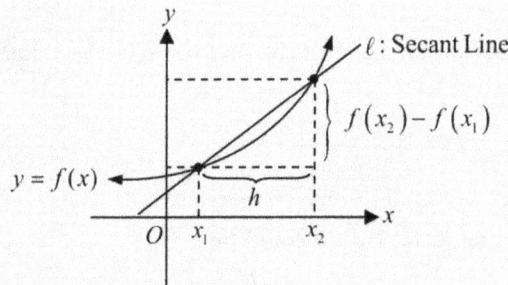

The instantaneous rate of change of y with respect to x at x_1 given by

$$r_{ins} = \lim_{x_2 \to x_1} \frac{f(x_2) - f(x_1)}{x_2 - x_1} = \lim_{h \to 0} \frac{f(x_1 + h) - f(x_1)}{h}$$

It also represents the slope (gradient) of the **tangent line** ℓ, which can also be represented by the gradient function $f'(x)$.

$$m_{\text{tangent}} = f'(x) = \lim_{h \to 0} \frac{f(x_1 + h) - f(x_1)}{h}$$

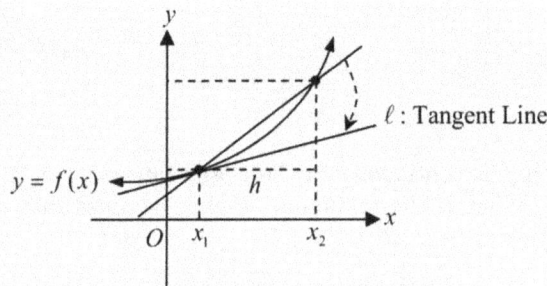

The *instantaneous rate of change* of a function is called the **derivative**.

Theorem 2.2 (Derivative of a Function).

The derivative of the function f with respect to the variable x is the function f' whose value at x is

$$f'(x) = \lim_{h \to 0} \frac{f(x + h) - f(x)}{h}, \quad \text{provided the limit exists.}$$

The derivative of the function f at the point $x = a$ is the limit

$$f'(a) = \lim_{h \to 0} \frac{f(a + h) - f(a)}{h}, \text{ or } \lim_{x \to a} \frac{f(x) - f(a)}{x - a}, \quad \text{provided the limit exists.}$$

The slope of the tangent line on the curve $y = f(x)$ at point $(a, f(a))$ is $f'(a)$

Claim 2.3 (Notation of Derivative).

Note that the **derivative** of $f(x)$ is given by the limit $\lim\limits_{h\to 0} \dfrac{f(x+h)-f(x)}{h}$. Ther are two common notations to express the derivative.

1. **Lagrange's notation** is to write the derivative of the function as $f'(x)$ or y'

   when $x = a$, the derivative is denoted as $f'(a)$ or $y'(a)$.

2. **Leibniz's notation** is to write the derivative of the function as $\dfrac{\mathrm{d}f}{\mathrm{d}x}$ or $\dfrac{\mathrm{d}}{\mathrm{d}x}f(x)$ or $\dfrac{\mathrm{d}y}{\mathrm{d}x}$

   when $x = a$, the derivative is denoted as $\left.\dfrac{\mathrm{d}f}{\mathrm{d}x}\right|_{x=a}$ or $\left.\dfrac{\mathrm{d}}{\mathrm{d}x}f(x)\right|_{x=a}$ or $\left.\dfrac{\mathrm{d}y}{\mathrm{d}x}\right|_{x=a}$.

Example (1). If $f(x) = x^2 - 2x - 3$, find: (a) $f'(x)$ using the definition of derivative, (b) $f'(0)$, (c) $f'(1)$, and (d) $f'(3)$.

Solution.

Example (2). Using the definition of derivative to find the derivative of $f(x) = \dfrac{1}{x-1}$.

Solution.

Example (3). Using the definition of derivative to find the derivative of a quadratic function $f(x) = ax^2 + bx + c$.

Solution.

Example (4). Evaluate the limit $\displaystyle\lim_{h\to 0} \dfrac{\cos(\pi + h) - \cos\pi}{h}$.

Solution.

2.2 Differentiability and Continuity

Before we start to learn the rules of differentiation, we need to focus on the features of function that is considered to be differentiable.

Theorem 2.4 (Differentiability of a Function).

A **continuous function** f is said to be differentiable at $x = a$ if the following limit exists.

$$\lim_{h \to 0} \frac{f(a+h) - f(a)}{h}$$

This is equivalent to say that the derivatives from the right and from the left are **identical**.

$$\lim_{h \to 0^+} \frac{f(a+h) - f(a)}{h} = \lim_{h \to 0^-} \frac{f(a+h) - f(a)}{h}$$

We may notice that the function need to be continuous when we discuss about differentiablity, thus we may duduce that:

Corollary 2.5 (Continuity vs. Differentiability).

If f is differentiable at $x = a$ then, f **must be** continuous at $x = a$.

If f is continuous at $x = a$ then, f **is not necessarily** differentiable at $x = a$.

Here are several examples of function that are not differentiable at a given number $x = a$.

(1) If $f(x)$ is discontinuous at $x = a$, then $f(x)$ is not differentiable at $x = a$

(2) The continuous function $f(x)$ does not have a finite derivative.

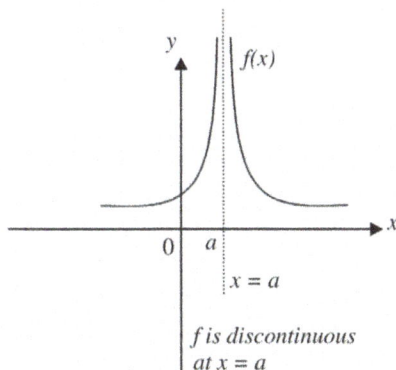

f is discontinuous at x = a

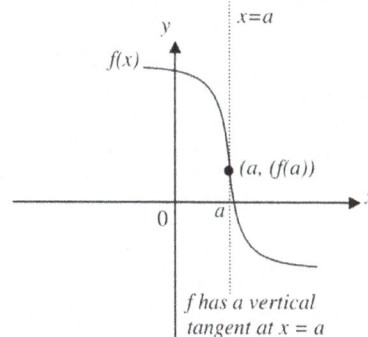

f has a vertical tangent at x = a

(3) The continuous function $f(x)$ **does not have identical derivatives** from the left and right.

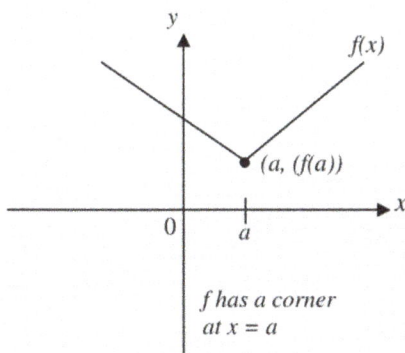

f has a corner at x = a

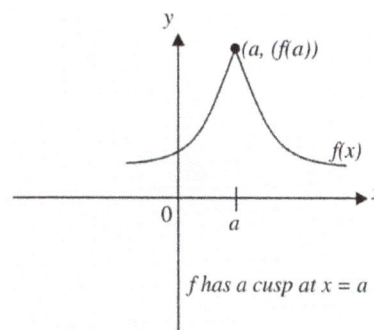

f has a cusp at x = a

In this case, we may find a *sharp corner or cusp* within the trajectory of the curve. However, the graph of a differentiable function should be a **continuous and smooth** curve.

Example (1). The Graph of a function is defined on the interval $[-6, 12]$. Find every -value at which the function is not differentiable on $(-6, 12)$.

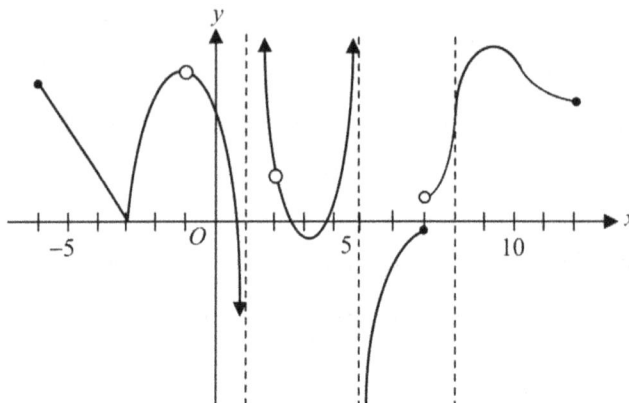

Solution.

Example (2). Determine continuity and differentiability of $f(x) = |2x|$.

Solution.

Example (3). Determine continuity and differentiability of $f(x) = x^2 - 3|x| + 2$.

Solution.

Example (4). Determine whether the function is differentiable at $x = 1$.

(a) $f(x) = \sqrt{1 - x^2}$

(b) $f(x) = \begin{cases} x + 1, & x \leq 1 \\ x^2 + 1, & x > 1 \end{cases}$

Solution.

2.3 Rules for Differentiation

The basic differentiation rules allow us to compute the derivatives of such functions without using the formal definition of the derivative. Consider these rules in more detail.

Theorem 2.6 (Basic Differentiation Rules).

1. **The Constant Rule:** The derivative of a constant function is 0.

$$\frac{\mathrm{d}}{\mathrm{d}x}[k] = 0, \text{ where } k \text{ is a real constant.}$$

2. **The Power Rule:** For any real number n:

$$\frac{\mathrm{d}}{\mathrm{d}x}[x^n] = nx^{n-1}$$

3. **Scalar Multiple Rule:** For any differentiable function f and g:

$$\frac{\mathrm{d}}{\mathrm{d}x}[kf(x)] = k\frac{\mathrm{d}}{\mathrm{d}x}[f(x)] = kf'(x)$$

4. **The Sum and Difference Rule:** for any function f and g:

$$\frac{\mathrm{d}}{\mathrm{d}x}[f(x) \pm g(x)] = \frac{\mathrm{d}}{\mathrm{d}x}[f(x)] \pm \frac{\mathrm{d}}{\mathrm{d}x}[g(x)]$$

5. **Linear Combination Rule:** For a, b as arbitrary constants:

$$\frac{\mathrm{d}}{\mathrm{d}x}[af(x) + bg(x)] = a \cdot \frac{\mathrm{d}}{\mathrm{d}x}[f(x)] + b \cdot \frac{\mathrm{d}}{\mathrm{d}x}[g(x)] = af'(x) + bg'(x)$$

Example (1). If $f(x) = 2x^3$, find $f'(x)$.

Solution.

Example (2). If $y = \dfrac{1}{x^2}$, find $\dfrac{\mathrm{d}y}{\mathrm{d}x}$ and $\dfrac{\mathrm{d}y}{\mathrm{d}x}\bigg|_{x=0}$.

Solution.

Example (3). Find $f'(x)$ and $f'(3)$ if $f(x) = \dfrac{1}{\sqrt{x}}$.

Solution.

Example (4). Find $f'(x)$ if $f(x) = x^3 - 10x + 5$.

Solution.

Theorem 2.7 (Product and Quotient Rule).

1. **The Product Rule:** For any differentiable function u and v.

$$\frac{\mathrm{d}}{\mathrm{d}x}[u(x)v(x)] = \frac{\mathrm{d}}{\mathrm{d}x}[u(x)] \cdot v(x) + \frac{\mathrm{d}}{\mathrm{d}x}[v(x)] \cdot u(x) = u'v + v'u$$

2. **The Quotient Rule:** For any real number n:

$$\frac{\mathrm{d}}{\mathrm{d}x}\left[\frac{u(x)}{v(x)}\right] = \frac{\frac{\mathrm{d}}{\mathrm{d}x}[u(x)] \cdot v(x) - \frac{\mathrm{d}}{\mathrm{d}x}[v(x)] \cdot u(x)}{[v(x)]^2} = \frac{u'v - v'u}{v^2}$$

Example (5). If $y = (3x - 5)(x^4 + 8x - 1)$, find $\dfrac{\mathrm{d}y}{\mathrm{d}x}$.

Solution.

Example (6). If $f(x) = \dfrac{2x - 1}{x + 5}$, find $f'(x)$.

Solution.

2.4   Derivatives of Exponential, Logarithmic and Trigonometric Functions

For the other functions, we also have certain formulae and rules to simplify our procedure to find the derivatives.

Theorem 2.8 (Derivatives of Other Functions).

1. **Derivatives of Exponential and Logarithmic Functions**

$$\frac{\mathrm{d}}{\mathrm{d}x}[e^x] = e^x \qquad\qquad \frac{\mathrm{d}}{\mathrm{d}x}[a^x] = a^x \cdot \ln a, \ (a > 0, a \neq 1)$$

$$\frac{\mathrm{d}}{\mathrm{d}x}[\ln x] = \frac{1}{x} \qquad\qquad \frac{\mathrm{d}}{\mathrm{d}x}[\log_a x] = \frac{1}{x \ln a}, \ (a > 0, a \neq 1)$$

2. **Derivatives of Exponential and Logarithmic Functions**

$$\frac{\mathrm{d}}{\mathrm{d}x}[\sin x] = \cos x \qquad \frac{\mathrm{d}}{\mathrm{d}x}[\cos x] = -\sin x \qquad \frac{\mathrm{d}}{\mathrm{d}x}[\tan x] = \sec^2 x$$

$$\frac{\mathrm{d}}{\mathrm{d}x}[\sec x] = \sec x \tan x \qquad \frac{\mathrm{d}}{\mathrm{d}x}[\csc x] = -\csc x \cot x \qquad \frac{\mathrm{d}}{\mathrm{d}x}[\cot x] = -\csc^2 x$$

Example (1). Use the *change-of-base formula* to show that the derivative of $y = \log_2 x$ is $\dfrac{1}{x \ln 2}$.

Solution.

Example (2). Use the quotient rule to show that the derivative of $y = \tan x$ is $\sec^2 x$.

Solution.

By using the **Quotient Rule** and setting the numerator as 1, we can deduce the formula for the reciprocal of any differential function f.

Corollary 2.9 (Derivative of Reciprocal Function).

If the derivative of $f(x)$ is $f'(x)$, then the derivative of $\dfrac{1}{f(x)}$ is:

$$\frac{\mathrm{d}}{\mathrm{d}x}\left[\frac{1}{f(x)}\right] = \frac{0 \cdot f(x) - f'(x) \cdot 1}{(f(x))^2} = -\frac{f'(x)}{(f(x))^2}$$

Example (3). If $f(x) = e^x \cos x$, find $f'(0)$.

Solution.

Example (4). If $f(x) = \dfrac{\sin x}{x^2}$, find $f'(x)$.

Solution.

Example (5). If $y = \dfrac{\tan x}{1 + \tan x}$, find $\dfrac{dy}{dx}$.

Solution.

Example (6). If $y = 5xe^3 + x^2 e^x$, find $\dfrac{dy}{dx}$.

Solution.

Example (7). If $y = \dfrac{\ln x}{e^x}$, find $\dfrac{dy}{dx}$.

Solution.

2.5 Tangent and Normal Line

In the previous section, we have discussed about the relationship between the derivative and the slope of tangent line. In fact, we may also use the derivative at certain point to find the equation of the tangent line.

Theorem 2.10 (Equation of Tangent and Normal Lines).

The equation of the **tangent line** of $y = f(x)$ through point (x_1, y_1) is

$$y - y_1 = m(x - x_1), \text{ where } m = f'(x_1).$$

The equation of the **normal line** of $y = f(x)$ through point (x_1, y_1) is

$$y - y_1 = -\frac{1}{m}(x - x_1), \text{ where } m = f'(x_1).$$

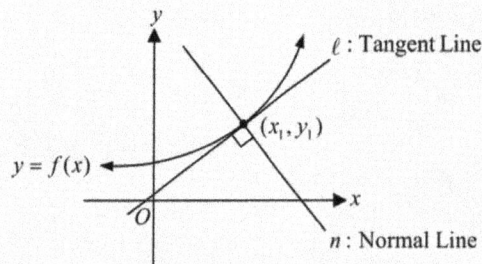

The normal line is defined as the line that is perpendicular to the tangent line at the point of tangency.

Example (1). Find the equation of the tangent line to the curve $y = x \ln x$ at $x = e$.

Solution.

Example (2). Write an equation for each normal to the graph of $y = 2 \sin x$ for $0 \le x \le 2\pi$ that has a slope of $\frac{1}{2}$.

Solution.

Example (3). Find all points on the graph of $y = |xe^x|$ at which the graph has a horizontal tangent.

Solution.

3   Differentiation: Composite, Implicit and Inverse functions

3.1   The Chain Rule

To compute the derivative of the composition of two or more functions, we may need the **chain rule**.

> **Theorem 3.1** (The Chain Rule).
>
> If $y = f(u)$ and $u = g(x)$ are differentiable functions of u and x respectively, then the derivative of the **composite function** $y = f(g(x))$ is denoted as:
>
> $$\frac{\mathrm{d}}{\mathrm{d}x}\left[f(g(x))\right] = f'(u) \cdot g'(x) = f'(g(x)) \cdot g'(x)$$
>
> It can also be expressed in the Leibniz's Form:
>
> $$\frac{\mathrm{d}y}{\mathrm{d}x} = \frac{\mathrm{d}y}{\mathrm{d}u} \times \frac{\mathrm{d}u}{\mathrm{d}x}$$

> **Example** (1). If $f(x) = (3x - 5)^{10}$, find $\dfrac{\mathrm{d}y}{\mathrm{d}x}$.

> **Solution.**

From the previous example, we may deduce a *special form* of chain rule to differentiate $f(ax + b)$.

> **Lemma 3.2.**
>
> For any differentiable function in the form $y = f(ax + b)$, if we let $u = ax + b$, then:
>
> $$\frac{\mathrm{d}y}{\mathrm{d}u} = f'(u), \qquad \frac{\mathrm{d}u}{\mathrm{d}x} = a.$$
>
> By the **chain rule**, we will deduce that:
>
> $$\frac{\mathrm{d}y}{\mathrm{d}x} = \frac{\mathrm{d}y}{\mathrm{d}u} \times \frac{\mathrm{d}u}{\mathrm{d}x} = a \cdot f'(u), \text{ where } u = ax + b.$$

For some *elementary function* we have discussed about in the previous unit, we may deduce that:

> **Corollary 3.3** (Derivative of $f(ax + b)$).
>
> The derivative of the following function is denoted as:
>
> $$\frac{\mathrm{d}}{\mathrm{d}x}(ax + b)^n = an(ax + b)^{n-1} \qquad \frac{\mathrm{d}}{\mathrm{d}x}\left(\frac{1}{ax + b}\right) = -\frac{a}{(ax + b)^2}$$
>
> $$\frac{\mathrm{d}}{\mathrm{d}x}e^{ax+b} = a \cdot e^{ax+b} \qquad \frac{\mathrm{d}}{\mathrm{d}x}\ln ax + b = \frac{a}{ax + b}$$
>
> $$\frac{\mathrm{d}}{\mathrm{d}x}\sin(ax + b) = a \cdot \cos(ax + b) \qquad \frac{\mathrm{d}}{\mathrm{d}x}\cos(ax + b) = -a \cdot \sin(ax + b)$$
>
> $$\frac{\mathrm{d}}{\mathrm{d}x}\tan(ax + b) = a \cdot \sec^2(ax + b) \qquad \frac{\mathrm{d}}{\mathrm{d}x}\cot(ax + b) = -a \cdot \csc^2(ax + b)$$
>
> $$\frac{\mathrm{d}}{\mathrm{d}x}\sec(ax + b) = a \cdot \sec(ax + b)\tan(ax + b) \qquad \frac{\mathrm{d}}{\mathrm{d}x}\csc(ax + b) = -a \cdot \csc(ax + b)\cot(ax + b)$$

Example (2). Find $f'(x)$ if $f(x) = \cot(4x - 6)$.

Solution.

Example (3). Find $f'(x)$ if $f(x) = e^{-2x}$.

Solution.

Example (4). Find $f'(x)$ if $f(x) = e^{x^3+3}$.

Solution.

Example (5). If $y = 3^{\sin x}$, find $\dfrac{dy}{dx}$.

Solution.

Example (6). If $y = (\ln x)^5$, find $\dfrac{dy}{dx}$.

Solution.

Example (7). If $y = \log_5\left(x^2 + 2x - 3\right)$, find $\dfrac{dy}{dx}$.

Solution.

Example (8). If $f(x) = 5x\sqrt{25 - x^2}$, find $f'(x)$.

Solution.

Example (9). If $g(x) = \csc^3(2x + 1)$, find $g'(x)$.

Solution.

Example (10). If $y = \sin\left(\cos\left(2x\right)\right)$, find $\dfrac{dy}{dx}$.

Solution.

3.2 Implicit Differentiation

So far, we have learnt how to differentiate $y = f(x)$, and theses functions are said to be given *explicitly*. In many problems, however, the function can be defined in **implicit form**, such as: $x^2 + y^2 = 9, x\cos y = y\ln xy$, etc.

Claim 3.4 (Method to Find Derivatives of Implicit Functions).

Given an equation containing the variables x and y for which you cannot easily solve for y in terms of x, you can find $\dfrac{dy}{dx}$ by doing the following:

Step 1: Differentiate each term of the equation with respect to x.

Step 2: Move all terms containing $\dfrac{dy}{dx}$ to the left side and all other terms to the right side.

Step 3: Factor out $\dfrac{dy}{dx}$ on the left side of the equation.

Step 4: Solve for $\dfrac{dy}{dx}$.

From the previous CLAIM, we may encounter the differentiation of some expressions involving y instead of x, here we may also use the **chain rule** to solve the problem.

Lemma 3.5. For any expression $f(y)$ involving y, the derivative with respect to x is:

$$\frac{d}{dx}\,f(y) = \frac{d}{dx}\,f(y) \times \frac{dy}{dx} = f'(y)\cdot\frac{dy}{dx}$$

Example (1). Find $\dfrac{dy}{dx}$ if $x^2 + y^2 = 9$.

Solution.

Example (2). Find $\dfrac{dy}{dx}$ if $y^2 - 7y + x^2 - 4x = 10$.

Solution.

In some complicated functions, we may need to differentiate $f(x)g(y)$ or $\dfrac{f(x)}{g(y)}$, here the **product and quotient rules** still apply for these cases.

Corollary 3.6 (Derivative of a Combination of x and y Expressions).

The derivatives of $f(x)g(y)$ and $\dfrac{f(x)}{g(y)}$ is denoted as:

$$\frac{d}{dx}[f(x)g(y)] = f'(x)g(y) + f(x)g'(y)\frac{dy}{dx}, \qquad \frac{d}{dx}\left[\frac{f(x)}{g(y)}\right] = \frac{f'(x)g(y) - f(x)g'(y)\frac{dy}{dx}}{[g(y)]^2}.$$

In particular, when $f(x) = x$ and $g(y) = y$

$$\frac{d}{dx}(xy) = y + x\frac{dy}{dx}, \qquad \frac{d}{dx}\left(\frac{x}{y}\right) = \frac{y - x\frac{dy}{dx}}{y^2} = \frac{1}{y} - \frac{x}{y^2}\cdot\frac{dy}{dx}.$$

Example (3). Find $\dfrac{dy}{dx}$ if $x^3 + y^3 = 6xy$.

Solution.

Example (4). Find $\dfrac{\mathrm{d}y}{\mathrm{d}x}$ if $x \sin y = \cos(x+y)$.

Solution.

Example (5). Where is the tangent to the curve $4x^2 + 9y^2 = 36$ vertical?

Solution.

3.3 Derivatives of Inverse Function

In some cases, we may need to determine the derivatives of the inverse functions (i.e. the inverse trigonometric function). Here we need the following technique:

Claim 3.7 (Slope of Inverse Function).

Point $A(a,b)$ and point $B(b,a)$ are corresponding points of f and f^{-1} respectively. The slope at point A is $\dfrac{dy}{dx}$ and the slope of at point B is $\dfrac{dx}{dy}$. Therefore, $\dfrac{dy}{dx} = \dfrac{1}{dx/dy}$

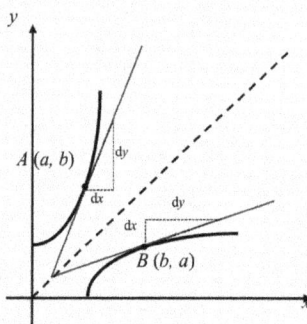

Then we may deduce that: $\left(f^{-1}\right)'(b) = \dfrac{1}{f'(a)}$, where $b = f(a)$ and $a = f^{-1}(b)$

We summarize this result in the following theorem.

Theorem 3.8 (Inverse Function Theorem).

Let $f(x)$ be a function that is both invertible and differentiable. Let $f^{-1}(x)$ be the inverse of $f(x)$. For all x satisfying $f^{-1}(x) \neq 0$:

$$\frac{d}{dx}\left(f^{-1}(x)\right) = \left(f^{-1}\right)'(x) = \frac{1}{f'\left(f^{-1}(x)\right)}$$

Alternatively, if $f^{-1}(x) = g(x)$ then:

$$g'(x) = \frac{1}{f'(g(x))} = \frac{1}{f'\left(f^{-1}(x)\right)}$$

Example (1). For $f(x) = e^x$, show that $\dfrac{d}{dx}(\ln x) = \dfrac{1}{x}$

Solution.

Sometimes, it is hard to find the inverse function $g(x)$ directly, then we might choose the latter two methods to solve the problem.

Example (2). Consider $f(x) = x^3 + 2x - 10$, if f^{-1} exists find $\left(f^{-1}\right)'(x)$

Solution.

From the previous example, we may also conclude that:

Corollary 3.9.

If f that has an inverse, then the derivative of $f^{-1}(x)$ can be expressed in terms of y:

$$\left(f^{-1}\right)'(x) = \frac{1}{f'(y)}$$

To avoid confusion, the y in the previous equation represents the y-coordinate of the inverse function instead of the original one. Therefore, it is algebraically equivalent to $f^{-1}(x)$, NOT $f(x)$.

In particular, for any $f^{-1}(x)$ passing through (b, a),

$$\left(f^{-1}\right)'(b) = \frac{1}{f'(a)} \quad \text{, where } a \text{ can be solved from } f(a) = b$$

Example (3). For $f(x) = x^5 + 3x - 8$, find $\left(f^{-1}\right)'(-8)$

Solution.

Example (4). For $f(x) = \cos x$ $(0 \leq x \leq \pi)$, find $\left(f^{-1}\right)'\left(\dfrac{1}{2}\right)$

Solution.

3.4   Derivatives of Inverse Trigonometric Functions

In this part, we are going to look at the derivatives of the inverse trigonometric functions based on the **inverse function theorem**.

Theorem 3.10 (Derivatives of Inverse Trigonometric Functions).

The derivatives of the inverse trigonometric functions are:

$$\frac{d}{dx}\left[\sin^{-1}x\right] = \frac{1}{\sqrt{1-x^2}} \qquad \frac{d}{dx}\left[\cos^{-1}x\right] = -\frac{1}{\sqrt{1-x^2}}$$

$$\frac{d}{dx}\left[\tan^{-1}x\right] = \frac{1}{1+x^2} \qquad \frac{d}{dx}\left[\cot^{-1}x\right] = -\frac{1}{1+x^2}$$

$$\frac{d}{dx}\left[\sec^{-1}x\right] = \frac{1}{|x|\sqrt{x^2-1}} \qquad \frac{d}{dx}\left[\csc^{-1}x\right] = -\frac{1}{|x|\sqrt{x^2-1}}$$

If $u = u(x)$ is a differentiable function of x, then:

$$\frac{d}{dx}\left[\sin^{-1}u(x)\right] = \frac{u'(x)}{\sqrt{1-(u(x))^2}} \qquad \frac{d}{dx}\left[\cos^{-1}u(x)\right] = -\frac{u'(x)}{\sqrt{1-(u(x))^2}}$$

$$\frac{d}{dx}\left[\tan^{-1}u(x)\right] = \frac{u'(x)}{1+(u(x))^2} \qquad \frac{d}{dx}\left[\cot^{-1}u(x)\right] = -\frac{u'(x)}{1+(u(x))^2}$$

$$\frac{d}{dx}\left[\sec^{-1}u(x)\right] = \frac{u'(x)}{|u(x)|\sqrt{(u(x))^2-1}} \qquad \frac{d}{dx}\left[\csc^{-1}u(x)\right] = -\frac{u'(x)}{|u(x)|\sqrt{(u(x))^2-1}}$$

Example (1). Find $g'(x)$ if $g(x) = \arcsin x + \sqrt{1-x^2}$

Solution.

Example (2). Find $f'(x)$ if $f(x) = \tan^{-1}\sqrt{x}$

Solution.

Example (3). If $y = \cos^{-1}\left(\dfrac{1}{x}\right)$, find $\dfrac{dy}{dx}$

Solution.

Example (4). If $\arcsin x = \ln y$, find $\dfrac{dy}{dx}$

3.5 Higher Order Derivatives

In some cases, we may need to differentiate a function for several times, the result is named as the n-th order derivatives, where n is the number of times we differentiate the function.

Theorem 3.11.

If the derivative f of a function f' is differentiable, then the derivative of f' is the second derivative of f represented by f'' (reads as f double prime). You can continue to differentiate f as long as there is differentiability.

Some of the symbols of Higher Order Derivatives:

a. Lagrange's Notation: $f'(x)$, $f''(x)$, $f'''(x)$, $f^{(4)}(x)$, ... $f^{(n)}(x)$

b. Leibniz's Notation: $\dfrac{dy}{dx}$, $\dfrac{d^2y}{dx^2}$, $\dfrac{d^3y}{dx^3}$, $\dfrac{d^4y}{dx^4}$, ... $\dfrac{d^ny}{dx^n}$

Example (1). If $f(x) = \sqrt{x}$, find $f''(4)$

Solution.

Example (2). If $y = x\cos x$, find $\dfrac{d^2y}{dx^2}$

Solution.

Example (3). If $x - y^2 = m$, find $\dfrac{d^2y}{dx^2}$ at the point where $y = -1$

Solution.

4 Contextual Applications of Differentiation

4.1 Interpreting the Meaning of the Derivative in Context

In Unit 2, we have defined the derivative of $f(x)$ at $x = a$ as $\lim_{h \to 0} \dfrac{f(a+h) - f(a)}{h}$. One application for derivatives is to find the **instantaneous rate of change** of a variable.

> **Theorem 4.1** (Average and Instanteous Rate of Change).
>
> The **average rate of change** of the function f over the interval $[a, a+h]$ or $[a, b]$ is the ratio of the amount of change of y over that interval to the corresponding change in the x values. It is given by:
>
> $$\text{Average Rate of Change } = \frac{f(b) - f(a)}{b - a} = \frac{f(a+h) - f(a)}{h}$$
>
> The **instantaneous rate of change** is defined as the limit when the change in the *independent variable x* approaches 0:
>
> $$\text{Instantaneous Rate of Change } = \lim_{b \to a} \frac{f(b) - f(a)}{b - a} = \lim_{h \to 0} \frac{f(a+h) - f(a)}{h} = f'(a)$$
>
> The unit for $f'(x)$ is equivalent to the unit for f divided by the unit for x.

Example (1). Let $G = 400(15 - t)^2$ be the number of gallons of water in a cistern t minutes after an outlet pipe is opened. Find the average rate of drainage during the first 5 minutes and the instantaneous rate at which the water is running out at the end of 5 minutes, and explain these rates in context.

Solution.

From the previous example we may deduce from the sign of the rate that:

> **Corollary 4.2.**
>
> When the rate of change is **positive**, the *dependent variable* is **increasing**.
>
> When the rate of change is **negative**, the *dependent variable* is **decreasing**.

Example (2). The function $t = f(M)$ models the time, in seconds, for a chemical reaction to occur as the mass M of a catalyst used, measured in grams. What is the unit for $f'(M)$

Solution.

4.2   Straight-Line Motion Problems

The motion problem is one of the most important applications in calculus. The techniques we have discussed earlier can help us to connect the position, velocity and acceleration of a moving object.

Theorem 4.3 (Instantaneous Velocity and Acceleration).

A particle moving along a straight has a position function $x(t)$. We can use the derivatives of $x(t)$ to find the velocity and acceleration.

1. Velocity function: $v(t) = x'(t) = \dfrac{\mathrm{d}x}{\mathrm{d}t}$

   If a particle is moving to the **right** (in positive direction), then $v(t) > 0$

   If a particle is moving to the **left** (in negative direction), then $v(t) < 0$

2. Acceleration function: $a(t) = v'(t) = \dfrac{\mathrm{d}v}{\mathrm{d}t} = x''(t) = \dfrac{\mathrm{d}^2 x}{\mathrm{d}t^2}$

3. Instantaneous speed: $|v(t)|$

Example (1). The position of a particle moving on a straight line is $x(t) = 2t^3 - 10t^2 + 5$. Find:

(a) the position at $t = 1$

(b) the instantaneous velocity at $t = 1$

(c) the acceleration at $t = 1$

(d) the speed of the particle at $t = 1$.

Solution.

Corollary 4.4 (Change in Speed and Velocity).

For any moving object with velocity $v(t)$ and acceleration $a(t)$.

1. If $v(t)$ and $a(t)$ have the **same sign**, the particle's **speed is increasing**.

2. If $v(t)$ and $a(t)$ have the **opposite sign**, the particle's **speed is decreasing**.

3. If $v(t) = 0$, the object comes to an **instantaneous rest**.

4. If $v(t) = 0$ and $v(t)$ **changes its sign**, the object changes its direction of moving.

In some cases, we need to be careful about the term **displacement** and **distance**.

Corollary 4.5 (Distance and Displacement).

The **displacement** of an moving object from t_1 to t_2 is: $x(t_2) - x(t_1)$

The **distance** of the object is $x(t_2) - x(t_1)$ only when the object *is not changing direction*. If it changes direction, we need to sum up all displacements in each part of the motion.

Example (2). A particle moves along the x-axis so that its position is $x(t) = -t - e^{1-t}$.

(a) Find the velocity function.

(b) Find the acceleration function.

(c) Is the speed of the particle increasing at time $t = 4$?

(d) Find all values of t at which the particle changes direction.

(e) Find the total distance traveled by the particle over the time interval $0 \leq t \leq 4$.

Solution.

Example (3). A particle moves along the x-axis so that its position is $x(t) = t^3 - 6t^2 + 9t + 3$.

(a) For $0 \leq t \leq 6$, find all times t during the particle is moving to the right.

(b) Find the acceleration at time $t = 4$. Is the speed of the particle increasing at $t = 4$?

(c) Find all times t in the open interval $0 < t < 6$ when particle changes direction.

(d) Find the total distance traveled by the particle from $t = 0$ until time $t = 6$.

(e) During $0 \leq t \leq 4$, what is the greatest distance between the particle from the origin?

Solution.

Example (4).

The graph of the velocity is show in the figure.

(a) When is the acceleration $= 0$?

(b) When is the particle moving to the right?

(c) When is the speed the greatest?

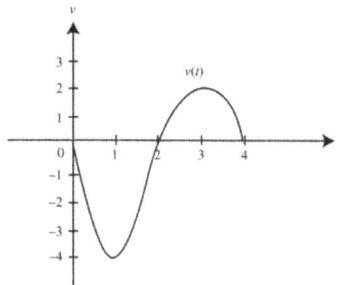

Solution.

Example (5). The graph of the position function of a moving particle is shown in the rogue below.

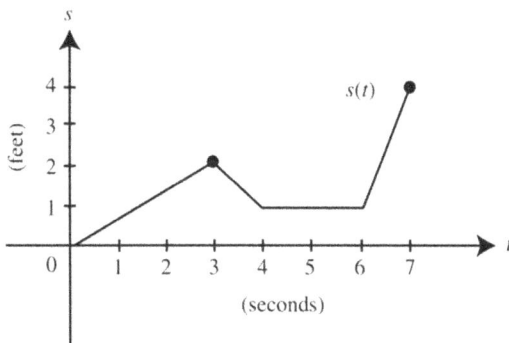

(a) What is the particle's position at $t = 5$?

(b) When is the particle moving to the left?

(c) When is the particle standing still?

(d) When does the particle have the greatest speed?

Solution.

4.3   Related Rates

Suppose we have two quantities, which are connected to each other and both changing with time. A related rates problem is a problem in which we know the rate of change of one of the quantities and want to find the rate of change of the other quantity.

Claim 4.6.

The way to solve related rates problems is as follows:

1. Identify the variables, including rates of change and the rate of change that is to be found.

2. Construct an equation relating the quantities whose rates of change are known to the quantity whose rate of change is to be found.

3. Implicitly differentiate both sides of the equation with respect to time.

4. Substitute the known rates of change and the known quantities into the equation.

5. Solve for the required rate of change.

Example (1). If x and y are both differentiable functions of t and are related by the equation $y = x^2 + 5$ and $\dfrac{dx}{dt} = 3$ when $x = 2$. Find $\dfrac{dy}{dt}$ when $x = 2$.

Solution.

Example (2). Find the surface area of a sphere at the instant when the rate of increase of the volume of the sphere is nine times the rate of increase of the radius.

Solution.

Example (3). A water tank has the shape of a cylinder with radius 5 meters. Let h be the depth of the water in the tank, measured in meters, where h is a function of time, t, measured in seconds. The volume V of the water tank is changing at the rate of -15π cubic meters per second. Find $\dfrac{dh}{dt}$.

Solution.

Example (4). Suppose that liquid is to be cleared of sediment by allowing it to drain through a conical filter that is 12 cm high and has a radius of 6 cm at the top. If the liquid is forced out of the cone at a constant rate of 2 cm^3/min, how fast is the depth of the liquid decreasing at the instant when the liquid in the cone is 4 cm deep?

Solution.

Example (5). A 25 feet- long ladder is leaning against the wall of a house and sliding away from the wall at a rate of 3 feet per second. How fast is the top of the ladder moving down the wall when the base of the ladder is 15 feet?

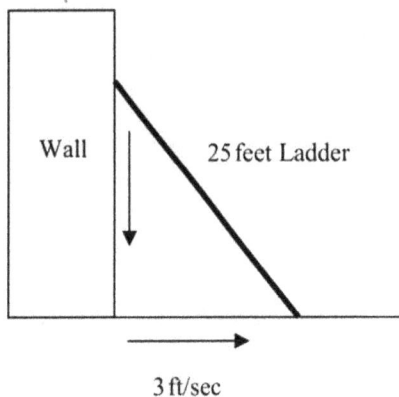

3 ft/sec

Solution.

4.4   Approximating Values Using Local Linearity and Linearization

In this section we're going to take a look at an application not of derivatives but of the tangent line to a function. Of course, to get the tangent line we do need to take derivatives, so in some way this is an application of derivatives as well.

Theorem 4.7 (Linear Approximation).

For a differentiable function f, the equation of the tangent line to its curve at the point $(a, f(a))$ is:

$$y = f'(a)(x - a) + f(a)$$

Since the curve of $f(x)$ and the tangent line are close to each other for points near $x = a$.

$$f(x) \approx f(a) + f'(a)(x - a)$$

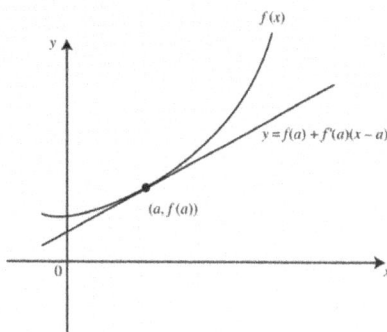

Example (1). Write an equation of the tangent line to $f(x) = x^3$ at $(2, 8)$. Use the tangent line to find the approximate values of $f(1.9)$ and $f(2.01)$.

Solution.

Example (2). Estimate the value of $\dfrac{3}{(1 - x)^2}$ at $x = 0.05$ from $x = 0$.

Solution.

From the previous examples, we may find that the approximation might be either less than or more than the actual value, it depends the curvature (or concavity) of the graph.

45

Corollary 4.8.

If the $f(x)$ is **concave up** within the interval containing a, then the linear approximation **under-estimates** the true value.

If the $f(x)$ is **concave down** within the interval containing a, then the linear approximation **over-estimates** the true value.

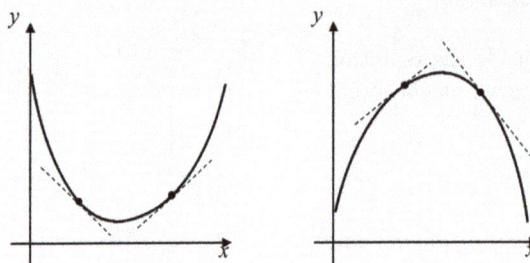

Example (3). Let $g(x) = e^x + 1$, and $g(0) = 2$ by considering the graph of function, find an approximation for $g(0.015)$ and state whether it is an overestimate or underestimate.

Solution.

46

4.5   Using L'Hospital's Rule for Finding Limits of Indeterminate Forms

The L'Hospital's Rule is an approach to find limits of indeterminate form (such as $\dfrac{0}{0}$, $\dfrac{\infty}{\infty}$ etc.), which is based on the differentiation of algebraic expressions.

Theorem 4.9 (L'Hospital's Rule).

Let f and g be differentiable on an open interval (a, b) containing c. If the limit of $\dfrac{f(x)}{g(x)}$ as x approaches c is indeterminate, then

$$\lim_{x \to c} \frac{f(x)}{g(x)} = \lim_{x \to c} \frac{f'(x)}{g'(x)}$$

The limit of $\dfrac{f(x)}{g(x)}$ is indeterminate if $\dfrac{f(x)}{g(x)} = \dfrac{0}{0}$, $\dfrac{\infty}{\infty}$, $0 \cdot \infty$, 0^0, 1^∞, ∞^0

Note: It is usually necessary to apply the rule more than once to remove indeterminate forms.

For different types of indeterminate forms, we have different approaches to find their limits.

Lemma 4.10.

If $\dfrac{f(x)}{g(x)} = \dfrac{0}{0}$, or $\dfrac{\infty}{\infty}$, then we can directly differentiate the numerator and denominator until the indeterminate form is removed.

Example (1). Show that $\lim\limits_{x \to 0} \dfrac{\sin x}{x} = 1$

Solution.

Example (2). Find $\lim\limits_{x \to 0} \dfrac{1 - \cos x}{x^2}$, if it exists

Solution.

Example (3). Find $\lim\limits_{x \to 0^+} \dfrac{\ln(x + 1)}{\sqrt{x}}$, if it exists

Solution.

Example (4). Find $\lim\limits_{x\to 0}\dfrac{e^x-1}{\tan 2x}$, if it exists

Solution.

Example (5). Find $\lim\limits_{x\to 0}\dfrac{2x^2}{e^{2x}-1}$, if it exists

Solution.

Lemma 4.11.

If $f(x)g(x) = 0 \cdot \pm\infty$, then we can change it to $\dfrac{f(x)}{1/g(x)}$ convert it to $\dfrac{0}{0}$, or $\dfrac{\infty}{\infty}$.

Example (6). Find $\lim\limits_{x\to\infty} x \cdot \sin\dfrac{1}{x}$, if it exists

Solution.

Example (7). Find $\lim\limits_{x\to -\infty} xe^x$, if it exists

Solution.

Lemma 4.12.

Other indeterminate forms, such as 0^0, 1^∞ and ∞^0, may be resolved by taking the natural logarithms and then use the L'Hospital's Rule.

Example (8). Find $\lim\limits_{x \to \infty} x^{\frac{1}{x}}$, if it exists

Solution.

Example (9). Find $\lim\limits_{x \to \infty} \left(1 + \dfrac{1}{x}\right)^{x}$, if it exists

Solution.

5 Contextual Applications of Differentiation

5.1 The Mean Value Theorem

The Mean Value Theorem is one of the most important theorems in calculus. We will look at some of its implications in this section.

Theorem 5.1 (Mean Value Theorem).

If f is a function that satisfies the following conditions:

1. f is continuous on the closed interval $[a, b]$.

2. f is differentiable on the open interval (a, b).

Then there is at least one point $x = c$ on this interval, such that $f'(c) = \dfrac{f(b) - f(a)}{b - a}$.

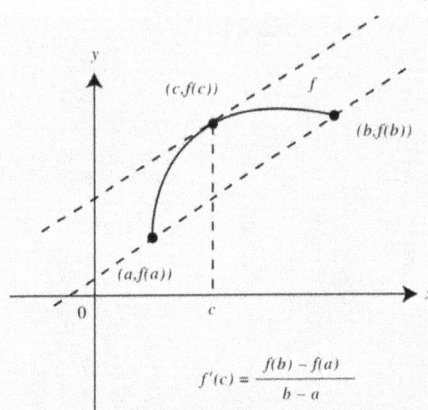

The geometric meaning of the **Mean Value Theorem** (MVT) ensures that there must be at least one point at which the tangent is parallel to the secant line joining $(a, f(a)$ and $(b, f(b))$.

Example (1). The function $f(x) = \dfrac{1}{4}x^3 + 1$ satisfies the **Mean Value Theorem** over the interval $[0, 2]$. Find all values of c in the interval $(0, 2)$ at which the tangent line to the graph of f is parallel to the secant line joining the points $(0, f(0))$ and $(2, f(2))$.

Solution.

Claim 5.2.

The contextual meaning of the MVT states that there must be at least one point $x = c$ on interval (a, b) such that the *instantaneous rate of change* at c is equal to the *average rate of change*.

Example (2). (Calculator required) The function f is defined by $f(x) = 3x - 4\cos(2x + 1)$. What are all values of x that satisfy the conclusion of the Mean Value Theorem applied to f on the interval $[-1, 2]$?

Solution.

A special case of MVT where $f(a) = f(b)$ is usually known as the **Rolle's Theorem**.

Corollary 5.3 (Rolle's Theorem).

Let f be a continuous function over $[a, b]$ and differentiable over (a, b).

If $f(a) = f(b)$. There then exists at least one $c \in (a, b)$ such that $f'(c) = 0$.

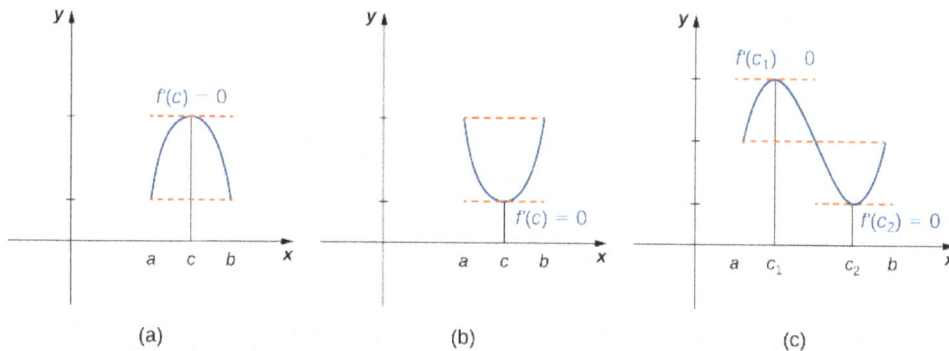

(a)   (b)   (c)

Geometrically speaking, there must exists at least one point at which the tangent is horizontal.

Example (3). Verify that the function $f(x) = 2x^2 - 8x + 6$ defined over the interval $[1, 3]$ satisfies the conditions of Rolle's theorem. Find all points c guaranteed by Rolle's theorem.

Solution.

5.2 Extreme Value Theorem and Critical Points

The derivative of a function can also help us determine the behaviour of functions, such as find the critical points, the trend and even the concavity of the curve.

Theorem 5.4 (Extreme Value Theorem).

If f is a continuous function on a closed interval $[a, b]$, then f has both a maximum and a minimum value on the interval.

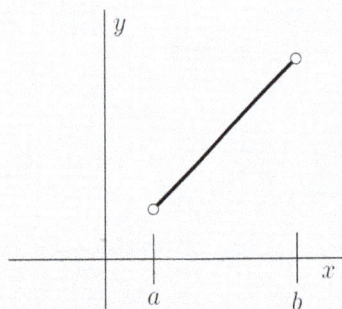

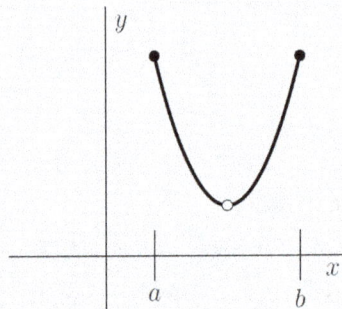

No max or min.   Two maxima, no mininmum.

The typical counterexamples are shown on the figure above, if f is not continuous at some points within the closed interval $[a, b]$, then the function might have no extrema.

To determine the extreme value, we need to find the **critical points** first.

Lemma 5.5 (Critical Point Theorem).

A **critical point** is an x-value $x = c$ in the domain of f such that $f'(c) = 0$ or $f'(c)$ is undefined.

The function $f(x)$ can only have extreme values at *critical points* or *endpoints*.

Example (1). Determine all critical points for $f(x) = \sqrt[3]{x^2}(2x + 1)$ within $\left[-\dfrac{1}{2}, \dfrac{1}{2} \right]$.

Solution.

Corollary 5.6 (Stationary Points vs. Critical Points).

A **stationary point** of a function $f(x)$ is a point where the derivative of $f(x)$ is equal to 0. These points are called "stationary" because the function is neither increasing nor decreasing. Note that if $f'(x)$ is undefined, the point is a critical point but not stationary point.

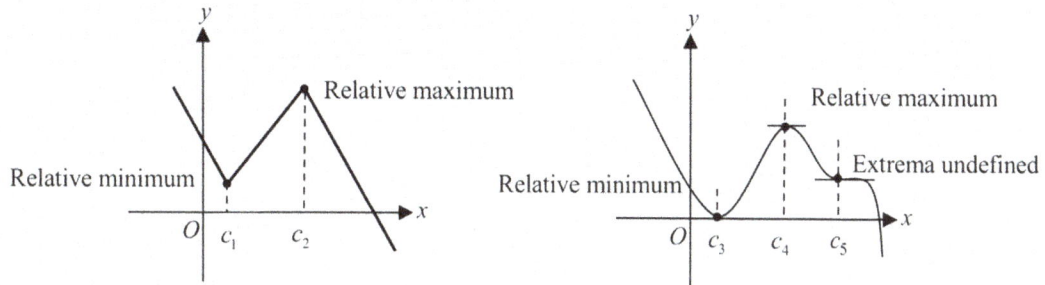

Points c_1 through c_5 are all **critical points**. critical points. c_3, c_4, and c_5 are also said to be **stationary points**.

Example (2). Determine all critical points and stationary points for $h(t) = 10t \cdot e^{3-t^2}$.

Solution.

5.3   Increasing and Decreasing Function

Theorem 5.7.

Let f be continuous on the closed interval $[a, b]$ and differentiable on the open interval (a, b).

1. If $f'(x) > 0$ on (a, b), then f is **increasing** on $[a, b]$.

2. If $f'(x) < 0$ on (a, b), then f is **decreasing** on $[a, b]$.

3. If $f'(x) = 0$ on (a, b), then f is **constant** on $[a, b]$.

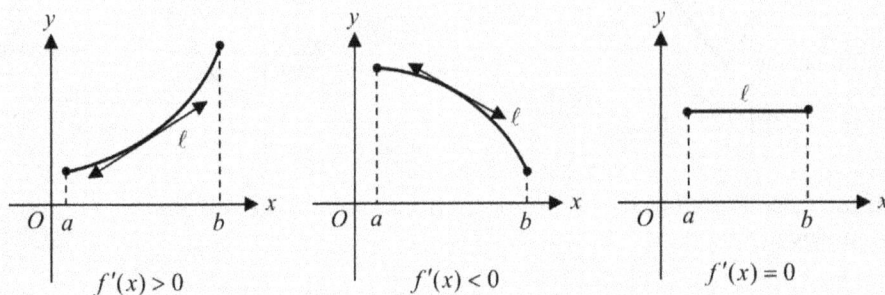

Example (1). Find the intervals on which $f(x) = x^3 - 3x^2$ is increasing or decreasing.

Solution.

Example (2). Find the intervals on which $f(x) = \dfrac{x^2}{x+1}$ is increasing or decreasing.

Solution.

5.4   Absolute and Relative Extrema

In this section, we will focus on the approaches to find the extrema of functions.

Theorem 5.8 (Absolute Extremum).

The extremum of a Function are the values of the maximum (or absolute maximum) and minimum (or absolute minimum) of a function on the interval $[a, b]$.

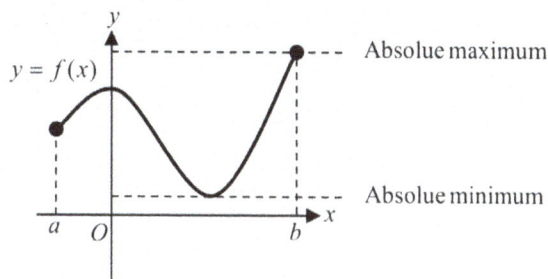

To find the absolute extrema, we need to look for some *candidates* including critical points and end points.

Theorem 5.9 (Relative Extremum).

Relative maximum (local maximum) is a maximum value relative to the points that are close to it on the graph.

Relative minimum (local minimum) is a minimum value relative to the points that are close to it on the graph.

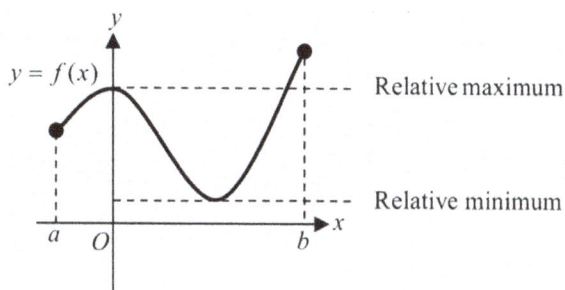

As is discussed in the previous section, the relative minimum and maximum can be determined by finding the critical points or stationary points via differentiating the function.

Corollary 5.10 (The First Derivative Test).

1. If $f'(x)$ changes from *negative* to *positive* at $x = k$, then $f(k)$ is a **relative minimum** of f.

2. If $f'(x)$ changes from *positive* to *negative* at $x = k$, then $f(k)$ is a **relative maximum** of f.

We can also clarify the type of stationary point using the **Second Derivative**.

Corollary 5.11 (The Second Derivative Test for Stationary Point).

1. If $f'(x_0) = 0$ and $f''(x_0) > 0$, then has a **relative minimum** at $x = x_0$.

2. If $f'(x_0) = 0$ and $f''(x_0) < 0$, then has a **relative maximum** at $x = x_0$.

3. If $f'(x_0) = 0$ and $f''(x_0) = 0$, the test is inconclusive at $x = x_0$.

Example (1). Find the relative extrema for the function $f(x) = \dfrac{x^3}{3} - x^2 - 3x$.

Solution.

Example (2). Find the relative maximum or relative minimum of $f(x) = \left(x^2 - 1\right)^{\frac{2}{3}}$.

Solution.

Example (3). Find the relative extrema of $f(x) = \dfrac{x^4 + 1}{x^2}$.

Solution.

Example (4). Show that the absolute minimum of $f(x) = \sqrt{25 - x^2}$ on $[-5, 5]$ is 0 and the absolute maximum is 5 .

Solution.

5.5   Determining Concavity of Functions

When studying a function, we may sometimes focus on the curvature of the graph, which is closely related to the change in the *rate of change*. The feature to represent the curvature is called **concavity**.

Theorem 5.12 (Concavity of Functions).

The Concavity of the function can be basically determined by:

1. If f is differentiable on an interval (a, b) and $f'(x)$ is increasing on the interval, the graph of $f(x)$ is **concave up** (upward).

2. If f is differentiable on an interval (a, b) and $f'(x)$ is decreasing on the interval, the graph of $f(x)$ is **concave down** (downward).

In addition, the second derivative $f''(x)$ can also help us to determine the concavity of the function.

1. If $f''(x) > 0$ for all x in interval (a, b), then the graph of $f(x)$ is **concave up**.

2. If $f''(x) < 0$ for all x in interval (a, b), then the graph of $f(x)$ is **concave down**.

3. If $f''(x) = 0$ for all x in interval (a, b), then the graph of $f(x)$ is **linear** (concavity undefined).

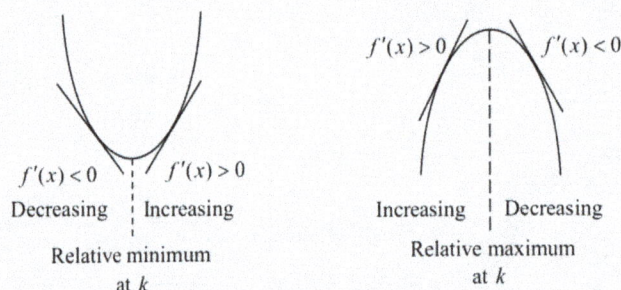

For some graphs of function, there might be partially concave up and partially concave down, then the point that separates these two types of curve are defined as follows:

Corollary 5.13 (Point of Inflection).

The **point of inflection** (inflection point) is a point on a curve at which the *concavity change*.

If f is a function that's twice differentiable almost everywhere except $x = c$, then the point $x = c$ of a curve $y = f(x)$ is a **point of inflection** if:

(a) $f''(c) = 0$ and f'' changes sign as x increases through c, or

(b) $f''(c)$ is undefined and f'' changes sign as x increases through c.

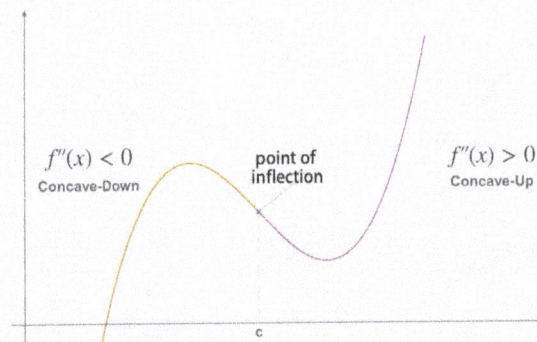

Example (1). Find the points of inflection of $f(x) = (x-1)^{\frac{1}{3}}$ and determine the intervals where the function f is concave up and where the function is concave down.

Solution.

Example (2). Given $f(x) = x + \sin x$, $(0 \le x \le 2\pi)$, find all points of inflection of f.

Solution.

Example (3). The graph of f is shown in the figure below and f is twice differentiable. Which of the following statements is true?

(A) $f(5) < f'(5) < f''(5)$

(B) $f''(5) < f'(5) < f(5)$

(C) $f'(5) < f(5) < f''(5)$

(D) $f'(5) < f''(5) < f(5)$

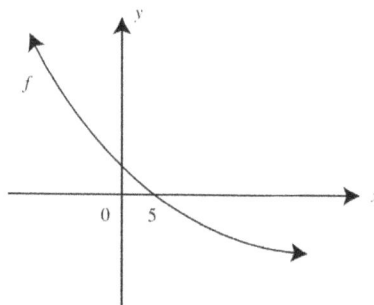

Solution.

Example (4).

The graph of f is shown in the figure below. Find the points of inflection of f and determine where the function f is concave upwards and where it is concave downwards on $[-3, 5]$.

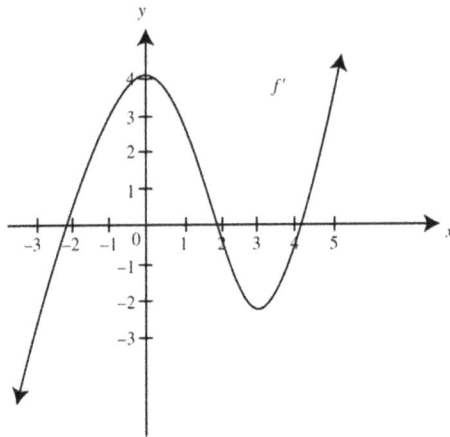

Solution.

5.6   Sketching Graphs of Functions

Claim 5.14.

When you are sketching the graph of a function, the following factors are useful.

1. Asymptotes

2. x- and y-intercepts

3. Critical points (These are found using the first derivative)

4. Points of inflection (These are found using the second derivative) and concavity.

Example (1).

Sketch the graph of $f(x) = x^3 - 3x^2 - 24x + 32$.

Solution.

Example (2).

Sketch the graph of $f(x) = \dfrac{x^2 - 3x + 2}{x}$.

Solution.

5.7   Connecting a Function, its First Derivative and Second Derivative

Example (1). Given the graph of f' in the figure below, find where the function f:

(a) has a horizontal tangent, (b) has its relative extrema, (c) is increasing or decreasing, (d) has a point of inflection, and (e) is concave upward or downward.

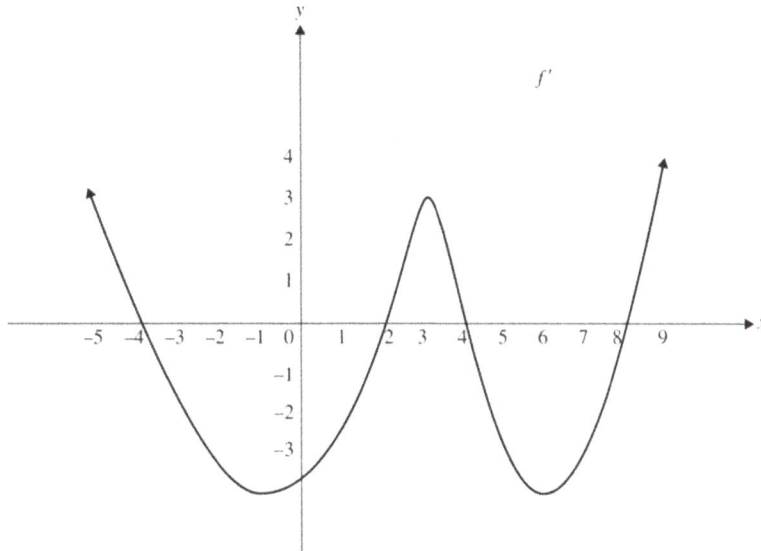

Solution.

Example (2). A function f is continuous on the interval $[-4, 3]$ with $f(-4) = 6$ and $f(3) = 2$ and the following properties:

INTERVALS	(−4, −2)	x = −2	(−2, 1)	x = 1	(1, 3)
f'	−	0	−	undefined	+
f''	+	0	−	undefined	−

(a) Find the intervals on which f is increasing or decreasing.

(b) Find where f has its absolute extrema.

(c) Find where f has the points of inflection.

(d) Find the intervals where f is concave upward or downward.

(e) Sketch a possible graph of f.

Solution.

Example (3). Give the graph of $f''(x)$ in the figure below, determine the values of x at which the function f has a point of inflection.

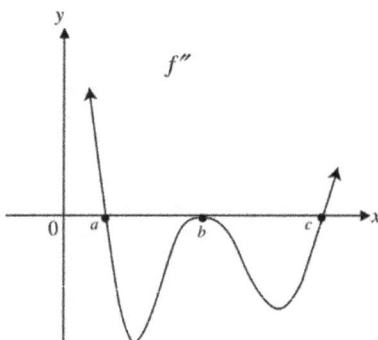

Solution.

Example (4). The graph of f is shown in the figure below and f is twice differentiable. Which of the following has the largest value:

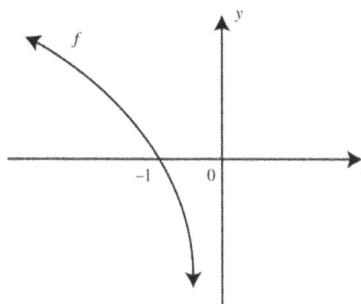

(A) $f(-1)$

(B) $f'(-1)$

(C) $f''(-1)$

(D) $f(-1)$ and $f'(-1)$

Solution.

5.8   Optimization Problems

The **optimization problem** is one of the most important problems involving calculus, we are looking for the largest value or the smallest value that a function can take. Which is exactly the same process to find the absolute extrema.

Claim 5.15.

Optimization is to find the largest value or the smallest value of a function subject to some kind of constraints (conditions)

Step 1: Define the primary equation for the quantity to be maximized.

Step 2: Identify the constraints.

Step 3: Reduce the primary equation to an equation with one variable.

Step 4: Use extrema to determine the desired maximum or minimum.

Example (1). Find the shortest distance between the point $A(19, 0)$ and the parabola $y = x^2 - 2x + 1$.

Solution.

Example (2). A window is constructed by adjoining a semicircle and a rectangle. If the total perimeter is 24 feet, what is the radius of the semicircle that will maximize the area of the window?

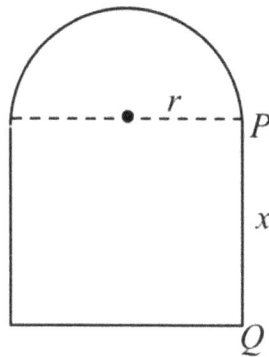

Solution.

Example (3). If an open box is to be made using a square sheet of tin, 20 inches by 20 inches, by cutting a square from each corner and folding the sides up, find the length of a side of the square being cut so that the box will have a maximum volume.

Solution.

6   Integration and Accumulation of Change

6.1   Exploring Accumulation of Change

The basic ideas of integral calculus is to find the **net change** (or *accumulated change*) of some changing variables. Which is usually considered as an inverse process of differentiation.

Theorem 6.1 (Net Change of Constant Rates).

If a dependent variable y is changing with a constant rate k with respect to x, then the net change is given by:

$$\Delta y = k \cdot \Delta x$$

which is graphically equivalent to the **rectangular area** under the horizontal line.

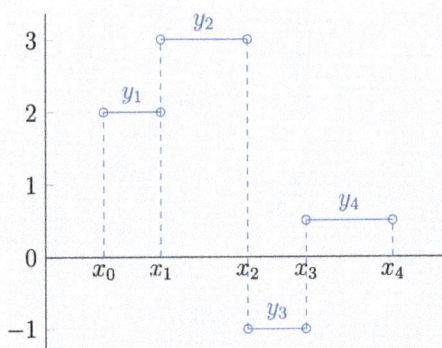

In the figure above, net changes of y with respect to x in the interval $[x_0, x_1]$, $[x_1, x_2]$, $[x_2, x_3]$ and $[x_3, x_4]$ are $2(x_1 - x_0)$, $3(x_2 - x_1)$, $(-1)(x_3 - x_2)$ and $\dfrac{1}{2}(x_4 - x_3)$ respectively. The accumulation of change from x_0 through x_4 is given by

$$2(x_1 - x_0) + 3(x_2 - x_1) + (-1)(x_3 - x_2) + \frac{1}{2}(x_4 - x_3)$$

which is also equivalent to the sum of the *signed* **rectangular areas**.

Note: if y has a negative value, the net change would be considered as a *negative value*.

For the accumulation of change with non-constant rate, we can still use the similar process to make the representation.

Corollary 6.2 (Net Change of Non-Constant Rates).

If a dependent variable y is changing with a non-constant rate with respect to x, then the accumulation of change is given by the **signed area under the curve**

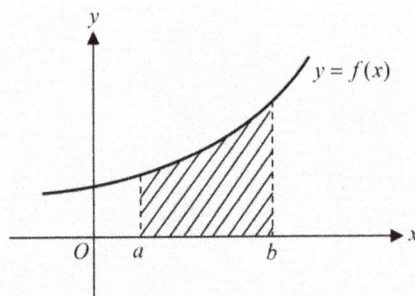

Which is usually denoted as $\displaystyle\int_a^b f(x)\mathrm{d}x$, and we will discussed the integral later.

The integration is used to determine the exact area and the accumulation of change of a variable y with respect to x within an interval, and we will discuss about the integral notation in the next few sections.

Example (1).

The graph shows the rate of change for the number of people in a museum t hours after it opens.

  (a) How many people are in the museum after 5 hours?

  (b) How many people are in the museum after 10 hours?

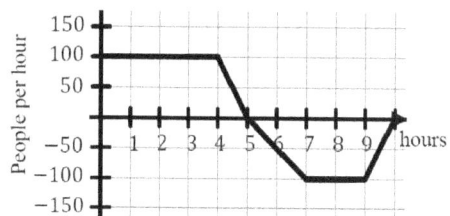

Solution.

Example (2).

The amount of water in a small pond is changing at the rate modeled in the graph, where the rate is measured in cubic inches per hour.

  (a) How much water has been lost during the first 3 hours?

  (b) How much water has been gained during the first 10 hours?

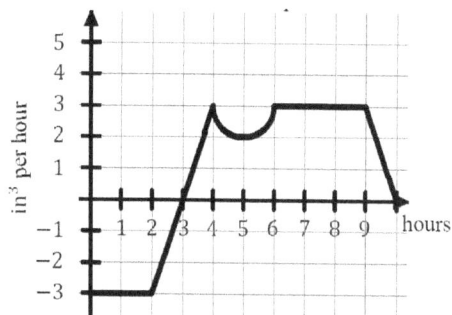

Solution.

6.2   Antiderivatives and Indefinite Integral

To find the exact area or accumulation of change for a function $f(x)$ whose graph is a curve, we need to know how to *integrate* things. As an inverse process of finding derivatives, the technique of taking integration is closely related to find the *antiderivative*.

Theorem 6.3 (Antiderivative).

A function F is an antiderivative of another function f if $F'(x) = f(x)$ for all x in some open interval. Any two antiderivatives of f differ by an additive constant C.

We denote the **set of antiderivatives** of f by $\int f(x)\ \mathrm{d}x$, called the indefinite integral of f.

The expression $f(x)$ is called the **integrand** and the variable x is the **variable of integration**.

The basic rules for indefinite integrals are:

1. $\displaystyle\int f(x)\ \mathrm{d}x = F(x) + C \ \Leftrightarrow\ F'(x) = f(x).$

2. $\displaystyle\int af(x)\ \mathrm{d}x = a\int f(x)\ \mathrm{d}x.$

3. $\displaystyle\int -f(x)\ \mathrm{d}x = -\int f(x)\ \mathrm{d}x.$

4. $\displaystyle\int f(x) \pm g(x)\ \mathrm{d}x = \int f(x)\ \mathrm{d}x \pm \int g(x)\ \mathrm{d}x.$

For some functions, evaluating indefinite integrals follows directly from properties of derivatives, such as the power functions and constants.

Corollary 6.4 (Constant Rule and Power Rule).

For any $n \neq -1$,

$$\int x^n\ \mathrm{d}x = \frac{x^{n+1}}{n+1} + C \qquad \int kx^n\ \mathrm{d}x = \frac{kx^{n+1}}{n+1} + C$$

In particular, when $n = 0$, the integrand becomes a constant k, then:

$$\int k\ \mathrm{d}x = kx + C$$

Example (1). Evaluate $\displaystyle\int x^5 - 6x^2 + x - 1\ \mathrm{d}x.$

Solution.

Example (2). Evaluate $\displaystyle\int 1 - \frac{1}{\sqrt[3]{x^4}}\ \mathrm{d}x.$

Solution.

Example (3). If $\dfrac{dy}{dx} = 3x^2 + 2$, and the point $(0, -1)$ lies on the graph of y, find y.

Solution.

Example (4). Evaluate $\displaystyle\int \dfrac{3x^3 + x^2 - 1}{x^2} \; dx$.

Solution.

Example (5). Evaluate $\displaystyle\int \sqrt{x} \left(x^2 - 3\right) \; dx$.

Solution.

In addition to the power functions, the other elementary function can also be integrated.

Corollary 6.5 (Antiderivatives of Trigonometric Functions).

Differentiation Formulas:

1. $\dfrac{d}{dx}(\cos x) = -\sin x$

2. $\dfrac{d}{dx}(\sin x) = \cos x$

3. $\dfrac{d}{dx}(\tan x) = \sec^2 x$

4. $\dfrac{d}{dx}(\cot x) = -\csc^2 x$

5. $\dfrac{d}{dx}(\sec x) = \sec x \tan x$

6. $\dfrac{d}{dx}(\csc x) = -\csc x \cot x$

Integration Formulas:

1. $\displaystyle\int \sin x \; dx = -\cos x + C$

2. $\displaystyle\int \cos x \; dx = \sin x + C$

3. $\displaystyle\int \sec^2 x \; dx = \tan x + C$

4. $\displaystyle\int \csc^2 x \; dx = -\cot x + C$

5. $\displaystyle\int \sec x \tan x \; dx = \sec x + C$

6. $\displaystyle\int \csc x \cot x \; dx = -\csc x + C$

Corollary 6.6 (Antiderivatives of Exponential,Logarithmic and Inverse Trigonometric Functions).

Differentiation Formulas:

1. $\dfrac{d}{dx}\left(e^x\right) = e^x$

2. $\dfrac{d}{dx}\left(a^x\right) = \ln a \cdot a^x$

3. $\dfrac{d}{dx}\left(\ln x\right) = \dfrac{1}{x}$

4. $\dfrac{d}{dx}\left(\sin^{-1} x\right) = \dfrac{1}{\sqrt{1-x^2}}$

5. $\dfrac{d}{dx}\left(\tan^{-1} x\right) = \dfrac{1}{1+x^2}$

6. $\dfrac{d}{dx}\left(\sec^{-1} x\right) = \dfrac{1}{|x|\sqrt{x^2-1}}$

Integration Formulas:

1. $\displaystyle\int e^x\,dx = e^x + C$

2. $\displaystyle\int a^x\,dx = \dfrac{a^x}{\ln a} + C$

3. $\displaystyle\int \dfrac{1}{x}\,dx = \ln|x| + C$

4. $\displaystyle\int \dfrac{1}{\sqrt{1-x^2}}\,dx = \sin^{-1} x + C$

5. $\displaystyle\int \dfrac{1}{1+x^2}\,dx = \tan^{-1} x + C$

6. $\displaystyle\int \dfrac{1}{|x|\sqrt{x^2-1}}\,dx = \sec^{-1} x + C$

Example (6). Evaluate $\displaystyle\int \dfrac{3x^2 + x - 1}{x^2}\,dx$.

Solution.

Example (7). Evaluate $\displaystyle\int x - \csc x \cot x \,dx$.

Solution.

Example (8). Evaluate $\displaystyle\int \sec x(\tan x - \sec x)\,dx$.

Solution.

Example (9). Evaluate $\displaystyle\int \tan^2 x \; dx$.

Solution.

Example (10). Evaluate $\displaystyle\int \frac{\sin x}{1 - \sin^2 x} \; dx$.

Solution.

Lemma 6.7. For any integrand in the form $f(ax)$ or $f(ax + b)$, if $\displaystyle\int f(x) \; dx = F(x) + C$, then:

$$\int f(ax) \; dx = \frac{1}{a} \cdot F(ax) + C, \qquad \int f(ax + b) \; dx = \frac{1}{a} \cdot F(ax + b) + C$$

Example (11). Evaluate $\displaystyle\int \frac{3}{2x + 1} \; dx$.

Solution.

Example (12). Evaluate $\displaystyle\int \frac{3}{e^x} \; dx$.

Solution.

Example (13). Evaluate $\displaystyle\int \sqrt[3]{(4x - 5)^2} \; dx$.

Solution.

Example (14). Evaluate $\displaystyle\int \sin^2 x \, \mathrm{d}x$.

Solution.

Example (15). Evaluate $\displaystyle\int \frac{1}{x^2 + 9} \, \mathrm{d}x$.

Solution.

6.3   Integration by U-Substitution

In this section we examine a technique, called integration by substitution, to help us find antiderivatives for more complicated functions. Specifically, this method helps us find antiderivatives when the integrand is the result of a chain-rule derivative.

Theorem 6.8 (Basic Substitution).

Recall that $\dfrac{\mathrm{d}}{\mathrm{d}x}\, F(g(x)) = F'(g(x)) \cdot g'(x)$ by the **chain rule,** the antiderivative can be written as:

$$\int F'(g(x))g'(x)\ \mathrm{d}x = F(g(x)) + C.$$

If $F'(x) = f(x)$, we can set the substitution $u = g(x)$, such that $\mathrm{d}u = g'(x)\ \mathrm{d}x$. Then:

$$\int f(g(x))g'(x)\ \mathrm{d}x = \int f(u)\ \mathrm{d}u = F(u) + C = F(g(x)) + C.$$

Basically, the U-substitution can be specified into 3 types.

Lemma 6.9 (Type I: integrating $(f(x))^n \cdot f'(x)$).

For $\displaystyle\int (f(x))^n \cdot f'(x)\ \mathrm{d}x$, $n \neq -1$, we may let $u = f(x)$ and $\mathrm{d}u = f'(x)\ \mathrm{d}x$, then:

$$\int (f(x))^n \cdot f'(x)\ \mathrm{d}x = \int u^n\ \mathrm{d}u = \frac{u^{n+1}}{n+1} + C = \frac{(f(x))^{n+1}}{n+1} + C.$$

Example (1). Evaluate $\displaystyle\int \left(x^2 + 4\right)^2 (2x)\ \mathrm{d}x$.

Solution.

Example (2). Evaluate $\displaystyle\int \sqrt{2x+1}\ \mathrm{d}x$.

Solution.

Example (3). Evaluate $\displaystyle\int \frac{x^2}{(x^3-8)^5}\ \mathrm{d}x.$

Solution.

Lemma 6.10 (Type II: integrating $\dfrac{f'(x)}{f(x)}$).

For $\displaystyle\int \frac{f'(x)}{f(x)}\ \mathrm{d}x$, we may let $u = f(x)$ and $\mathrm{d}u = f'(x)\ \mathrm{d}x$, then:

$$\int \frac{f'(x)}{f(x)}\ \mathrm{d}x = \int \frac{1}{u}\ \mathrm{d}u = \ln|u| + C = \ln|f(x)| + C.$$

Example (4). Evaluate $\displaystyle\int \frac{4x+6}{x^2+3x}\ \mathrm{d}x.$

Solution.

Example (5). Evaluate $\displaystyle\int \tan x\ \mathrm{d}x.$

Solution.

Example (6). By considering $\sec x = \dfrac{\sec x(\sec x + \tan x)}{\sec x + \tan x}$, evaluate $\displaystyle\int \sec x\ \mathrm{d}x.$

Solution.

From the examples from the previous two sections, we may go one step further about the trigonometric integrals.

> **Corollary 6.11** (Indefinite Integrals for Trigonometric Functions).
>
> The indefinite integrals for trigonometric functions and their squares are:
>
> $$\int \sin x \; \mathrm{d}x = -\cos x + C \qquad\qquad \int \sin^2 x \; \mathrm{d}x = \frac{1}{2}x - \frac{1}{4}\sin 2x + C$$
>
> $$\int \cos x \; \mathrm{d}x = \sin x + C \qquad\qquad \int \cos^2 x \; \mathrm{d}x = \frac{1}{2}x + \frac{1}{4}\sin 2x + C$$
>
> $$\int \tan x \; \mathrm{d}x = \ln|\sec x| + C \qquad\qquad \int \tan^2 x \; \mathrm{d}x = \tan x - x + C$$
>
> $$\int \cot x \; \mathrm{d}x = \ln|\sin x| + C \qquad\qquad \int \cot^2 x \; \mathrm{d}x = -\cot x - x + C$$
>
> $$\int \sec x \; \mathrm{d}x = \ln|\sec x + \tan x| + C \qquad\qquad \int \sec^2 x \; \mathrm{d}x = \tan x + C$$
>
> $$\int \csc x \; \mathrm{d}x = \ln|\csc x - \cot x| + C \qquad\qquad \int \csc^2 x \; \mathrm{d}x = -\cot x + C$$
>
> All of them can be resolved by trigonometric identities and u-substitutions.

For some other types of functions involving $f(g(x))$, we may also try to use the substitution $u = g(x)$ to determine the indefinite integrals.

Example (7). Evaluate $\int x\sqrt{2x+1} \; \mathrm{d}x$.

Solution.

Example (8). Evaluate $\int 2x^2 \cos\left(x^3\right) \; \mathrm{d}x$.

Solution.

Example (9). Evaluate $\int \sin^2 3x \cdot \cos 3x \ dx$.

Solution.

Example (10). Evaluate $\int x e^{3x^2} \ dx$.

Solution.

Example (11). Evaluate $\int \dfrac{1}{x \ln(x^2)} \ dx$.

Solution.

Example (12). Evaluate $\int \dfrac{\ln x}{3x} \ dx$.

Solution.

Example (13). Evaluate $\int \dfrac{1}{\sqrt{9-4x^2}}\ \mathrm{d}x$.

Solution.

Example (14). Evaluate $\int \dfrac{x(x-2)}{(x-1)^3}\ \mathrm{d}x$.

Solution.

6.4   Approximating Areas with Riemann Sums

Riemann Sum is a useful tool too determine the signed area that represents the accumulation of change for a variable, which is obtained by adding up the areas of multiple simplified slices of the region.

Theorem 6.12 (Definition Riemann Sum).

Let $f(x)$ be a continuous function defined on $[a,b]$ and x_i be points of $[a,b]$ such that

$$a = x_0 < x_1 < x_2 < x_3 \cdots < x_{n-1} < x_n = b$$

These points divide the interval $[a,b]$ into n subintervals:

$$[x_0, x_1], [x_1, x_2], \cdots, [x_{n-1}, x_n].$$

Let $\Delta x_i = x_i - x_{k-1}$ be the width of the k-th interval $[x_{k-1}, x_k]$ and c_k be any point within the k-th interval. The the **Riemann sum** of function f for the partition is given by:

$$\sum_{k=1}^{n} f(c_k)\Delta x_k = f(c_1)\Delta x_1 + f(c_2)\Delta x_2 + f(c_3)\Delta x_3 + ... + f(c_n)\Delta x_n$$

In particular, there are several types of Riemann Sums that can approximate the area bounded by the curve.

Theorem 6.13 (Three Types of Riemann Sum).

If f, the area A under the curve of f can be approximated using three common types of rectangles: left-endpoint rectangles, right-endpoint rectangles, or midpoint rectangles, with the same width $\Delta x = \dfrac{b-a}{n}$.

1. The Left Riemann Sum uses the left endpoints of the subintervals.

2. The Right Riemann Sum uses the right endpoints.

3. The Midpoint Riemann Sum is calculated using the midpoints of the subintervals.

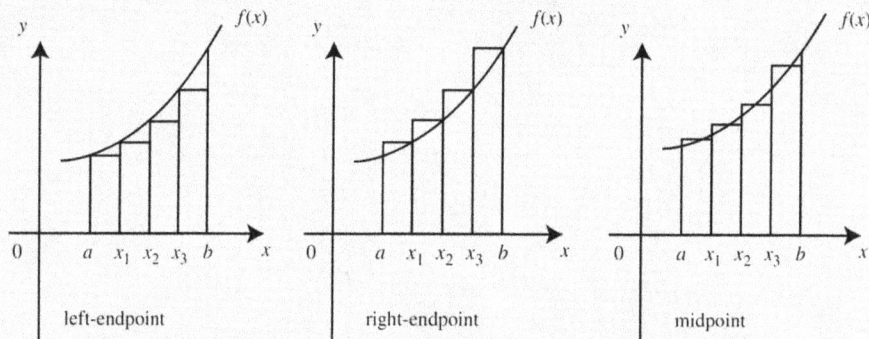

left-endpoint   right-endpoint   midpoint

The approximation of the total area bounded by the curve is given by

$$\text{Area} \approx \begin{cases} \sum_{k=1}^{n} f(x_{k-1})\Delta x, & \text{Left Riemann Sum} \\ \sum_{k=1}^{n} f(x_k)\Delta x, & \text{Right Riemann Sum} \\ \sum_{k=1}^{n} f\left(\dfrac{x_{k-1}+x_k}{2}\right)\Delta x, & \text{Minpoint Riemann Sum} \end{cases}$$

where $\Delta x = \dfrac{b-a}{n}$ and $a = x_0 < x_1 < x_2 < x_3... < x_{n-1} < x_n = b$.

Example (1). Find the approximate area under the curve of $f(x) = \sqrt{x}$ from $x = 4$ to $x = 9$ using 5 right-endpoint rectangles, and determine whether it is an overestimate or underestimate.

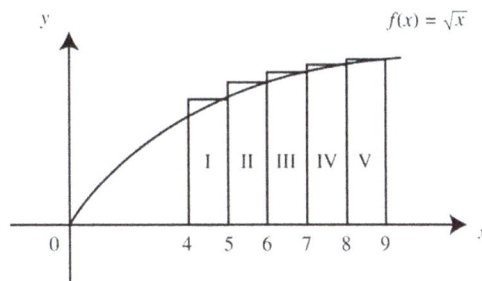

Solution.

Corollary 6.14.

Let $f(x)$ be a continuous and non-negative function defined on the closed interval $[a, b]$.

1. If $f(x)$ is increasing, then the left Riemann sum underestimates the area and right Riemann sum overestimates the area.

2. If $f(x)$ is decreasing, then the left Riemann sum overestimates the area and right Riemann sum underestimates the area.

Example (2). The function f is positive and continuous on $[1, 9]$. Selected values of f are given:

x	1	2	3	4	5	6	7	8	9
$f(x)$	1	1.41	1.73	2	2.37	2.45	2.65	2.83	3

Using 4 midpoint rectangles, approximate the area under the curve of f for $x = 1$ to $x = 9$.

Solution.

81

Another way to approximate area is to use the trapezoids.

Theorem 6.15 (Trapezoidal Approximations).

Let $f(x)$ be continuous on $[a, b]$. We partition the interval $[a, b]$ into n equal subintervals, each of width $\dfrac{b - a}{n}$ such that $a = x_0 < x_1 < x_2 < \cdots < x_n = b$.

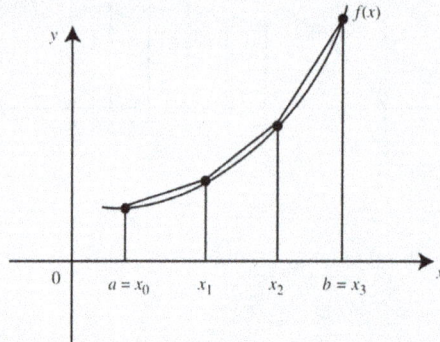

The trapezoidal approximation of area under the curve from $x = a$ to $x = b$ is:

$$\text{Area} \approx \frac{b - a}{2n} \left[f(x_0) + 2f(x_1) + 2f(x_2) + \cdots 2f(x_{n-1}) + f(x_n) \right]$$

Example (3). Find the approximate area under the curve of from to , using 4 trapezoids.

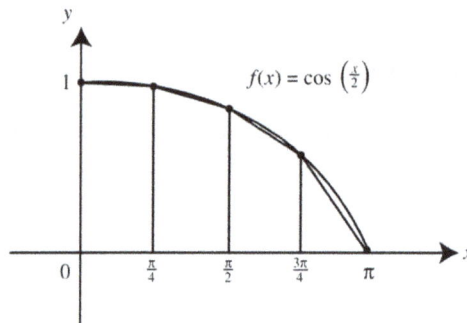

Solution.

Corollary 6.16.

Let $f(x)$ be a continuous and non-negative function defined on the closed interval $[a, b]$.

1. If $f(x)$ is concave up, then the trapezoidal approximation overestimates the area.

2. If $f(x)$ is concave down, then the trapezoidal approximation underestimates the area.

6.5   Riemann Sums, Summation Notation and Definite Integral Notation

In the previous section, we have introduced the Riemann sum to approximate the area under the curve via making partitions. If the length of each partition approaches 0, then the Riemann sum will also approaches the exact area. Then, we can use the limit of Riemann sum to express the idea.

Theorem 6.17 (Area and Definite Integral).

If we take the limit of the Riemann Sum as the width of the partition Δx approaches zero, we get the exact value of the area.

$$A = \lim_{\Delta x \to 0} \sum_{k=1}^{n} f(c_k)\Delta x_k.$$

The limit is called the definite integral of $f(x)$ from a to b and is denoted by: $\int_a^b f(x)\,\mathrm{d}x$

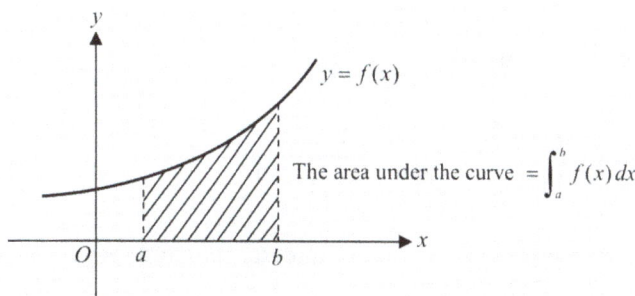

The notation for the definite integral is very similar to the notation for an indefinite integral. The new elements a and b mean, respectively, the lower and the upper limit of integration.

To clarify the conversion between the summation notation and integral notation, we usually make all the partitions of the interval $[a, b]$ having identical widths (i.e. $\Delta x = \dfrac{b-a}{n}$).

Corollary 6.18 (The Summation and Integral Notation).

The area under the curve of $f(x)$ on the interval $[a, b]$ is represented by:

$$\lim_{n \to \infty} \sum_{k=1}^{n} f\left(a + \frac{b-a}{n}k\right) \cdot \left(\frac{b-a}{n}\right) = \int_a^b f(x)\,\mathrm{d}x$$

Example (1). Write the summation notation for the integral $\int_1^7 \sqrt{x}\,\mathrm{d}x$ and $\int_0^6 \sqrt{1+x}\,\mathrm{d}x$

Solution.

Example (2). Write the integral notation for $\displaystyle\lim_{n\to\infty}\sum_{k=1}^{n}\cos\left(4+\frac{6k}{n}\right)\left(\frac{6}{n}\right)$, which lower limit $= 4$

Solution.

Example (3). Consider the expression

$$\lim_{n\to\infty}\left(\frac{2}{n}\right)\left(\frac{1}{\frac{2}{n}+3}+\frac{1}{\frac{4}{n}+3}+\frac{1}{\frac{6}{n}+3}+\cdots+\frac{1}{\frac{2n}{n}+3}\right)$$

Assuming the lower limit a is 0, write a definite integral that represents the above expression.

Solution.

Example (3). Which of the following integrals are equal to $\displaystyle\lim_{n\to\infty}\sum_{k=1}^{n}\left(-1+\frac{4k}{n}\right)^2\frac{4}{n}$

I. $\displaystyle\int_{-1}^{3} x^2 \,\mathrm{d}x$ II. $\displaystyle\int_{0}^{4} (-1+x)^2 \,\mathrm{d}x$ III. $\displaystyle\int_{0}^{1} 4(-1+4x)^2 \,\mathrm{d}x$

Solution.

6.6   The Fundamental Theorem of Calculus

earlier, the Fundamental Theorem of Calculus (FTC) is an extremely powerful theorem that establishes the relationship between differentiation and integration, and gives us a way to evaluate definite integrals without using Riemann sums or calculating areas.

Theorem 6.19 (The Fundamental Theorem of Calculus: Part 2).

If f is a continuous function on $[a, b]$, and F is an antiderivative of f (i.e. $F'(x) = f(x)$), then

$$\int_a^b f(x) \, \mathrm{d}x = F(x)\big|_a^b = F(b) - F(a)$$

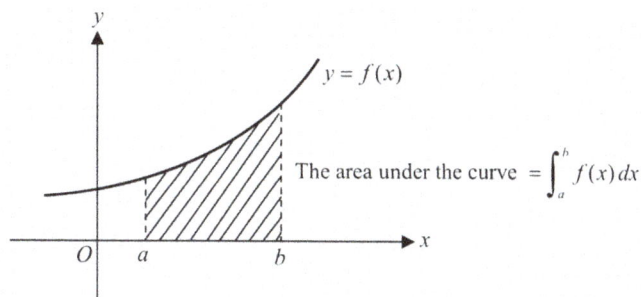

and this value is exactly the same as the **signed area** bounded by the curve of $f(x)$, x-axis, line $x = a$ and $x = b$.

From the previous theorem, we may deduce the following properties.

Corollary 6.20 (Properties for Definite Integrals).

For any definite integral for $f(x)$ from $x = a$ to $x = b$, we have:

1. $\displaystyle \int_a^a f(x) \, \mathrm{d}x = 0$

2. $\displaystyle \int_a^b f(x) \, \mathrm{d}x = \int_a^c f(x) \, \mathrm{d}x + \int_c^b f(x) \, \mathrm{d}x$ for $c \in (a, b)$

3. $\displaystyle \int_a^b f(x) \, \mathrm{d}x = - \int_b^a f(x) \, \mathrm{d}x$

4. $\displaystyle \int_{-a}^a f(x) \, \mathrm{d}x = 2 \int_0^a f(x) \, \mathrm{d}x$, where $f(x)$ is an *even function*.

5. $\displaystyle \int_{-a}^a f(x) \, \mathrm{d}x = 0$, where $f(x)$ is an *odd function*.

Example (1). Evaluate $\displaystyle \int_1^4 \frac{x^3 - 8}{\sqrt{x}} \, \mathrm{d}x$.

Solution.

Example (2). Evaluate $\displaystyle\int_{\frac{\pi}{4}}^{\frac{\pi}{2}} \csc^2 3t \ dt$.

Solution.

Corollary 6.21 (Using FTC with U-Substitution).

Let F and g be two differentiable functions at $[a, b]$, and $f(x) = F'(x)$, by the u-substitution, we have:
$$\int_a^b f(g(x)) \cdot g'(x) \ dx = \int_{g(a)}^{g(b)} f(u) \ du$$
By letting $u = g(x)$, the lower limit changes from a to $g(a)$, the upper limit changes from b to $g(b)$.

Example (3). Evaluate $\displaystyle\int_0^{\sqrt{\ln 2}} xe^{-x^2} \ dx$.

Solution.

Example (4). Evaluate $\displaystyle\int_1^{\sqrt[4]{3}} \frac{x}{1 + x^4} \ dx$.

Solution.

In some cases, we may need to find the definite integral for piecewise functions, then we need to deal with different expression for each of the subintervals.

Lemma 6.22. For any piecewise defined function:

$$f(x) = \begin{cases} f_1(x), & a \leq x \leq m \\ f_2(x), & m < x \leq b \end{cases}$$

The definite integral is given by: $\displaystyle \int_a^b f(x)\,\mathrm{d}x = \int_a^m f_1(x)\,\mathrm{d}x + \int_m^b f_2(x)\,\mathrm{d}x$

Example (5). Evaluate $\displaystyle \int_1^4 |3x - 6|\,\mathrm{d}x$.

Solution.

Theorem 6.23 (The Fundamental Theorem of Calculus: Part 1).

Let f be a continuous function on $[a, b]$. If the function g defined by

$$F(x) = \int_a^x f(t)\,\mathrm{d}t, \ a \leq x \leq b$$

is an antiderivative of f, then

$$F'(x) = f(x) \quad \text{or} \quad \frac{\mathrm{d}}{\mathrm{d}x}\left(\int_a^x f(t)\,\mathrm{d}t\right) = f(x)$$

If we substitute the upper limit with $u(x)$, then the **chain rule** applies:

$$\frac{\mathrm{d}}{\mathrm{d}x}\left(\int_a^{u(x)} f(t)\,\mathrm{d}t\right) = f(u(x)) \cdot u'(x)$$

Example (6). If $h(x) = \displaystyle \int_3^x \sqrt{t + 1}\,\mathrm{d}t$ find $h'(8)$

Solution.

Example (7). Find $\dfrac{\mathrm{d}}{\mathrm{d}x} \displaystyle\int_0^{\cos^2(2x)} \sqrt{t}\,\mathrm{d}t$

Solution.

Example (8). Find $\dfrac{\mathrm{d}y}{\mathrm{d}x}$ if $y = \displaystyle\int_{x^2}^1 \sin t\,\mathrm{d}t$

Solution.

6.7   Integrating Using Integration by Parts

Integration by Parts is a special method for integrating products of functions. Which can be usually considered as a reverse process of the **product rule** in differentiation.

> **Theorem 6.24** (Integration by Parts).
>
> If u and v are functions of x and differentiable, then
>
> $$\int u \, \mathrm{d}v = uv - \int v \, \mathrm{d}u$$
>
> It can also be expressed as the integration with respect to x:
>
> $$\int u(x) \cdot v'(x) \, \mathrm{d}x = uv - \int u'(x) \cdot v(x) \, \mathrm{d}x$$
>
> The key thing in integration by parts is to choose u and v' (or dv) correctly. The acronym (**ILATE**) is good for picking u, which stands for:
>
> Inverse Trigonometric – **L**ogarithmic – **A**lgebraic – **T**rigonometric – **E**xponential
>
> The closer a function is to the front, the more likely that it should be used as u.

> **Example** (1). Evaluate $\displaystyle\int xe^{-x} \, \mathrm{d}x$.
>
> Solution.

> **Example** (2). Evaluate $\displaystyle\int x^2 \ln x \, \mathrm{d}x$.
>
> Solution.

> **Example** (3). Evaluate $\displaystyle\int \ln x \, \mathrm{d}x$.
>
> Solution.

Example (4). Evaluate $\displaystyle\int e^x \cos x \, dx$.

Solution.

Example (5). By considering $\sec^3 x = \sec x \cdot \sec^2 x$, evaluate $\displaystyle\int \sec^3 x \, dx$.

Solution.

Lemma 6.25 (Tabular Integration by Parts).

When you have a problem involving repeated integration by parts, the Tabular Method is a nifty way to simplify repeated integration by parts especially for:

$$\int (\text{polynomial} \cdot \text{exponential}) \, dx \quad \text{or} \quad \int (\text{polynomial} \cdot \text{trigonometric}) \, dx$$

For example: $\int x^3 e^x \, dx$, we can use the following method.

Alternate Sign	x^3 and its derivative of x^3	e^x and its antiderivative of e^x	Product
$+$ →	x^3	e^x	
$-$ →	$3x^2$	e^x	$x^3 e^x$
$+$ →	$6x$	e^x	$-3x^2 e^x$
$-$ →	6	e^x	$6x e^x$
$+$	0	e^x	$-6e^x$

Therefore, $\int x^3 e^x \, dx = x^3 e^x - 3x^2 e^x + 6x e^x - 6e^x + C$.

Example (6). Evaluate $\int x^2 \sin x \, dx$.

Solution.

6.8   Integration Using Partial Fractions

Rational fractions can be integrated by splitting into **partial fractions**, which helps us to decompose a fraction to simpler forms (usually with linear or quadratic denominators).

Consider the addition of two rational fractions:

$$\frac{2}{x+1} + \frac{3}{x+4} = \frac{2(x+4) + 3(x+1)}{(x+1)(x+4)} = \frac{5x+11}{x^2+5x+4}.$$

Partial Fraction Decomposition is the *reverse of the procedure*. It allows to decompose a single rational function into a sum of simpler rational functions, for example

$$\frac{5x+11}{x^2+5x+4} = \frac{2}{x+1} + \frac{3}{x+4}.$$

The indefinite integral for the simplest rational functions is given by the following LEMMA.

> **Lemma 6.26.** For any rational function f with constant numerator and linear denominator, we have:
> $$\int \frac{A}{ax+b}\, \mathrm{d}x = \frac{A}{a}\ln|ax+b| + C.$$

To start with the partial fraction, we need to make sure:

> **Claim 6.27.** The method involves the following basic steps:
>
> 1. Check to make sure the fraction is a **proper rational function**.
>
> 2. Factor the polynomial in the denominator into linear and/or quadratic factors.
>
> 3. Decompose the fraction into a sum of partial fractions.

There are several cases to decompose rational functions into partial fractions.

> **Theorem 6.28** (Case 1: Denominator with Distinct Linear Factors).
>
> If a denominator of a fraction has two distinct linear factors, the fraction can be split into:
> $$\frac{px+q}{(ax+b)(cx+d)} = \frac{A}{ax+b} + \frac{B}{cx+d}$$
>
> The integration of this function is given by:
> $$\int \frac{px+q}{(ax+b)(cx+d)}\, \mathrm{d}x = \int \frac{A}{ax+b} + \frac{B}{cx+d}\, \mathrm{d}x$$
> $$= \frac{A}{a}\ln|ax+b| + \frac{B}{c}\ln|cx+d| + C.$$
>
> The rule can be extended to 3 or more linear factors.
>
> $$\frac{p(x)}{(a_1x+b_1)(a_2x+b_2)\cdots(a_nx+b_x)} = \frac{A_1}{a_1x+b_1} + \frac{A_2}{a_2x+b_2} + \cdots + \frac{A_n}{a_nx+b_1}$$
>
> And the integration of this function can be figured out in the same way.
>
> $$\int \frac{p(x)}{(a_1x+b_1)(a_2x+b_2)\cdots(a_nx+b_x)}\, \mathrm{d}x = \int \sum_{k=1}^{n} \frac{A_k}{a_kx+b_k}\, \mathrm{d}x$$
> $$= \sum_{k=1}^{n} \frac{A_k}{a_k}\ln|a_kx+b_k| + C.$$

Example (1). Evaluate $\displaystyle\int \frac{1}{x^2 + 3x - 4} \, dx$.

Solution.

Example (2). Evaluate $\displaystyle\int \frac{5x^2 + 7x}{(x^2 - 1)(2x + 1)} \, dx$.

Solution.

Theorem 6.29 (Case 2: Denominator with Repeating Linear Factors).

If a denominator of a fraction has repeating linear factors, the fraction can be split into:

$$\frac{px + q}{(ax + b)^2} = \frac{A}{ax + b} + \frac{B}{(ax + b)^2}$$

The integration of this function is given by:

$$\int \frac{px + q}{(ax + b)^2} \, dx = \int \frac{A}{ax + b} + \frac{B}{(ax + b)^2} \, dx$$
$$= \frac{A}{a} \ln|ax + b| - \frac{B}{a(ax + b)} + C.$$

Corollary 6.30 (Extended Cases for Repeating Linear Fraction).

If a denominator of a fraction contains repeating linear factors, the fraction can be split into:

$$\frac{px^2 + qx + r}{(ax + b)^2(cx + d)} = \frac{A}{ax + b} + \frac{B}{(ax + b)^2} + \frac{D}{cx + d}$$

The integration of this function is given by:

$$\int \frac{px^2 + qx + r}{(ax + b)^2(cx + d)} \, \mathrm{d}x = \int \frac{A}{ax + b} + \frac{B}{(ax + b)^2} + \frac{D}{cx + d} \, \mathrm{d}x$$
$$= \frac{A}{a} \ln|ax + b| - \frac{B}{a(ax + b)} + \frac{D}{c} \ln|cx + d| + C.$$

Example (3). Evaluate $\displaystyle\int \frac{x + 1}{(x - 1)^2} \, \mathrm{d}x$.

Solution.

Example (4). Evaluate $\displaystyle\int \frac{x + 1}{(x - 1)(x - 2)^2} \, \mathrm{d}x$.

Solution.

Theorem 6.31 (Case 3: Denominator with form $(ax+b)(cx^2+d)$).

If a denominator of a fraction has contains $(ax+b)(cx^2+d)$, the fraction can be split into:

$$\frac{px^2+qx+r}{(ax+b)(cx^2+d)} = \frac{A}{ax+b} + \frac{Bx+D}{cx^2+d} = \frac{A}{ax+b} + \frac{Bx}{cx^2+d} + \frac{D}{cx^2+d}$$

The integration of this function is given by:

$$\int \frac{px+q}{(ax+b)^2}\,dx = \int \frac{A}{ax+b} + \frac{Bx}{cx^2+d} + \frac{D}{cx^2+d}\,dx$$

$$= \frac{A}{a}\ln|ax+b| + \frac{B}{2c}\ln\left|cx^2+d\right| + \frac{D}{\sqrt{cd}}\tan^{-1}\left(\sqrt{\frac{c}{d}}x\right) + C.$$

Example (5). Evaluate $\displaystyle\int \frac{2x^2+x+1}{(x-1)(x^2+1)}\,dx$.

Solution.

Example (6). Evaluate $\displaystyle\int \frac{3-x}{(x+1)(x^2+3)}\,dx$.

Solution.

Theorem 6.32 (Integrating Improper Fractions).

For the improper fraction integrands, we need to reduce to the proper form and then use the partial fractions.

$$\int \frac{P(x)}{Q(x)} \, dx = \int F(x) + \frac{R(x)}{Q(x)} \, dx \quad \text{where} P(x) = F(x)Q(x) + R(x)$$

Example (7). Evaluate $\int \frac{2x^2 + 5x - 11}{x^2 + 2x - 3} \, dx$.

Solution.

Theorem 6.33 (Integration by Completing the Square).

By changing the square, we may rewrite any quadratic polynomial $ax^2 + bx + c$ in the form

$$a\left(x + \frac{b}{2a}\right)^2 + \frac{4ac - b^2}{4a^2}$$

If $b^2 - 4ac < 0$, the quadratic cannot be factorized, thus by completing the square:

$$\int \frac{1}{ax^2 + bx + c} \, dx = \int \frac{1}{a(x+m)^2 + k^2} \, dx = \frac{1}{k^2} \int \frac{1}{\frac{a(x+m)^2}{k^2} + 1} \, dx$$

where $m = \frac{b}{2a}$ and $k = \frac{\sqrt{4ac - b^2}}{2a}$, then we can use $\tan u = \frac{\sqrt{a}(x+m)}{k}$ to integrate.

Example (7). Evaluate $\int \frac{1}{x^2 + 2x + 5} \, dx$.

Solution.

6.9   Evaluating Improper Integrals

Sometimes, we need to find a definite integral that has either or both limits infinite or an integrand that approaches infinity at one or more points in the range of integration, these integrals are called **improper integrals**.

> **Theorem 6.34** (Improper Integrals).
>
> For $\int_a^b f(x)\,\mathrm{d}x$, there are two types of **improper integrals**.
>
> 1. The limit a or b (or both) are infinite;
>
> 2. The function $f(x)$ has one or more points of discontinuity in the interval $[a, b]$.

To evaluate the improper integrals, we need to apply the idea of **limits**.

> **Lemma 6.35** (Integration over an Infinite Domain).
>
> Let f be a continuous function for all real number x, then we consider the limit form:
>
> 1. $\displaystyle \int_a^\infty f(x)\,\mathrm{d}x = \lim_{b\to\infty} \int_a^b f(x)\,\mathrm{d}x = \lim_{b\to\infty} F(b) - F(a)$
>
> 2. $\displaystyle \int_{-\infty}^b f(x)\,\mathrm{d}x = \lim_{a\to-\infty} \int_a^b f(x)\,\mathrm{d}x = \lim_{a\to-\infty} F(b) - F(a)$
>
> 3. $\displaystyle \int_{-\infty}^\infty f(x)\,\mathrm{d}x = \int_{-\infty}^c f(x)\,\mathrm{d}x + \int_c^\infty f(x)\,\mathrm{d}x$
>
> where $F(x)$ is an antiderivative of $f(x)$.

Example (1). Evaluate $\displaystyle \int_1^\infty \frac{1}{x}\,\mathrm{d}x$.

Solution.

Example (2). Evaluate $\displaystyle \int_0^\infty \frac{1}{x^2 + 1}\,\mathrm{d}x$.

Solution.

Example (3). Evaluate $\int_{-\infty}^{\infty} xe^{-x^2} \, dx$.

Solution.

Lemma 6.36 (Integration with Infinite Discontinuities).

Let f be a function which is continuous on the interval $[a, b)$ but is discontinuous at $x = b$, then

$$\int_a^b f(x) \, dx = \lim_{k \to b^-} \int_a^k f(x) \, dx = \lim_{k \to b^-} (F(k) - F(a))$$

Let f be a function which is continuous on the interval $(a, b]$ but is discontinuous at $x = a$, then

$$\int_a^b f(x) \, dx = \lim_{k \to a^+} \int_k^b f(x) \, dx = \lim_{k \to a^+} (F(b) - F(k))$$

Let f be a function which is continuous on the interval $[a, b]$ except at $x = c$, then

$$\int_a^b f(x) \, dx = \int_a^c f(x) \, dx + \int_c^b f(x) \, dx$$

where $F(x)$ is an antiderivative of $f(x)$.

Example (4). Evaluate $\int_0^{\frac{\pi}{2}} \frac{\cos x}{\sqrt{1 - \sin x}} \, dx$.

Solution.

7 Differential Equations

7.1 Simple Differential Equations

In this unit, we are going to study the **differential equations**, the equations that contain any forms of **derivatives**.There is one differential equation that everybody probably knows, that is *Newton's Second Law of Motion*.

$$F_{net} = m \cdot \frac{\mathrm{d}v}{\mathrm{d}t}$$

There are many types of differential equations modelled from contexts, and the solution to those equation varies based on its type. The simplest equation is as follows:

Theorem 7.1 (Simple Differential Equations).

A simple differential equations contains only a derivative and its expression:

$$\frac{\mathrm{d}y}{\mathrm{d}x} = f(x)$$

We can solve these equations by finding **indefinite integrals**.

$$y = \int f(x) \, \mathrm{d}x = F(x) + C$$

The solution contains an **arbitrary constant** C is usually called the **general solution**.

To determine the specific value of C, we need an initial condition $y(a) = b$.

The solution contains a **specific constant** is usually called the **particular solution**.

Example (1). Find the general solution of the differentiation equation $\dfrac{\mathrm{d}v}{\mathrm{d}t} = 2t - e^{-t}$.

Solution.

Example (2). Find a solution of the differentiation equation $\dfrac{\mathrm{d}y}{\mathrm{d}x} = x \sin\left(x^2\right)$, with $y(0) = -1$.

Solution.

Example (3). Find the general solution of the differentiation equation $\dfrac{\mathrm{d}^2y}{\mathrm{d}x^2} = \dfrac{1}{x}$ with $x > 0$.

Solution.

Example (4). If $\dfrac{\mathrm{d}^2y}{\mathrm{d}x^2} = 2x + 1$ and at $x = 0$, $y' = -1$ and $y = 3$, find a solution of the differential equation.

Solution.

7.2   Sketching and Using Slope Field

Sometimes we would meet a differential equations not solvable by simple integration, such as $\dfrac{dy}{dx} = x + y$. We may need to use the **slope field** to help us find the solution curve.

Theorem 7.2 (Slope Field).

The derivative at a point of any differential equations gives a slope on this point.

$$\text{If } \frac{dy}{dx} = f(x, y), \text{ then slope } = y' = f(x, y)$$

A slope field is a graphical representation of the solutions of a first-order differential equation. It is achieved without solving the differential equation analytically, and thus it is useful. The representation may be used to qualitatively visualize solutions, or to numerically approximate them.

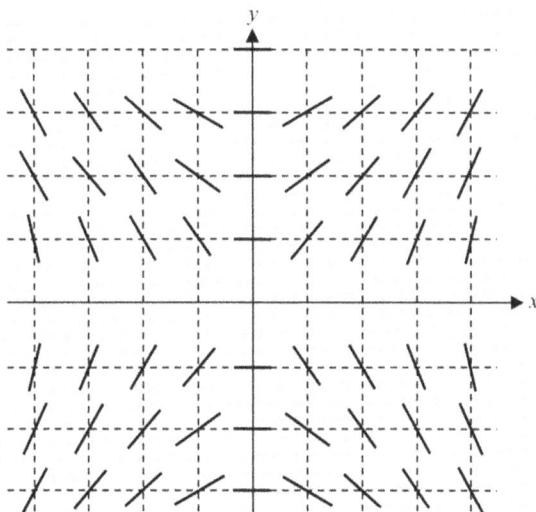

eg. $\dfrac{dy}{dx} = \dfrac{x}{y}$

Example (1). Sketch the slope field of the differential equation $\dfrac{dy}{dx} = x^2(y - 2)$ on the coordinate plane with vertices points with $-1 \le x \le 1$ and $0 \le y \le 5$.

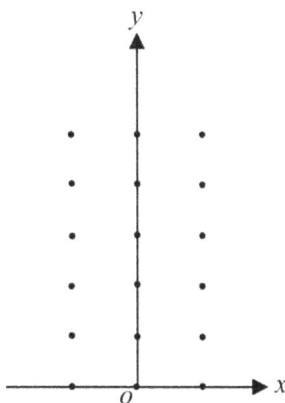

Solution.

Example (2). The figure below shows a slope field for one of the differential equations given below. Which of the following equation does the figure indicate.

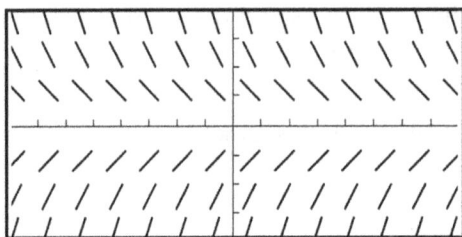

(A) $\dfrac{dy}{dx} = 2x$ (B) $\dfrac{dy}{dx} = -2x$ (C) $\dfrac{dy}{dx} = y$ (D) $\dfrac{dy}{dx} = -y$ (E) $\dfrac{dy}{dx} = x + y$

Solution.

Example (3). Which of the following could be a slope field for the differential equation $\dfrac{dy}{dx} = x^2 + y$?

(A)

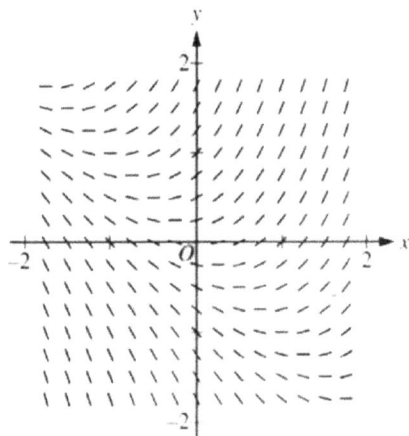

(B)

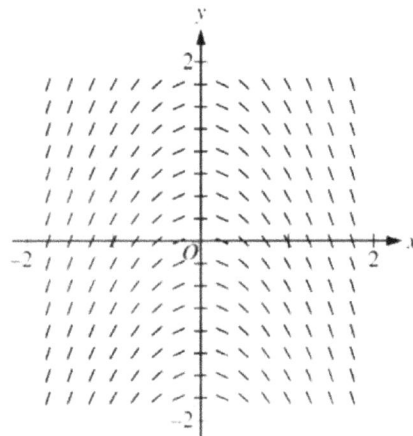

(C)

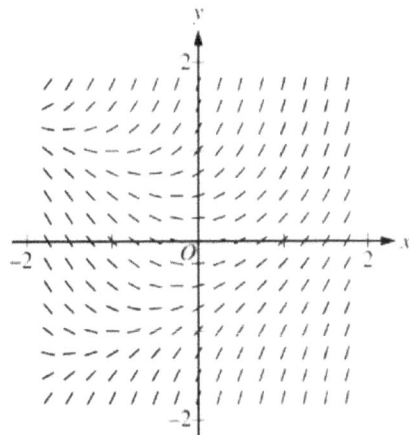

(D)

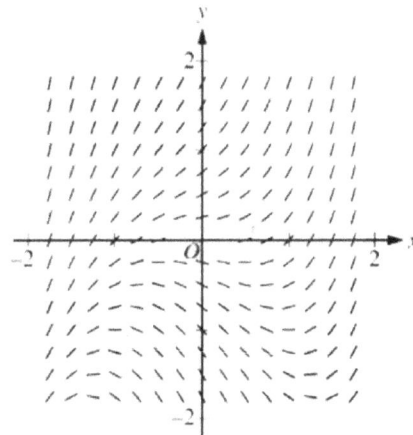

Solution.

Example (4). Match each slope field in the figures below with the proper differential equation from the following set. The particular solution that goes through $(0,0)$ has been sketched in.

(1) $\dfrac{dy}{dx} = \cos x$　　　(2) $\dfrac{dy}{dx} = 2x$　　　(3) $\dfrac{dy}{dx} = 3x^2 - 3$　　　(4) $\dfrac{dy}{dx} = -\dfrac{\pi}{2}$

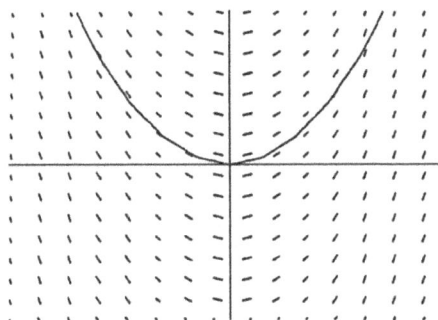

$[-2,2] \times [-2,2]$

(A)

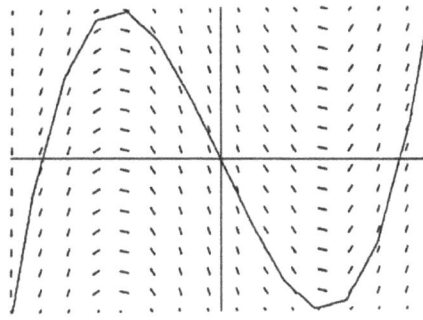

$[-2,2] \times [-2,2]$

(B)

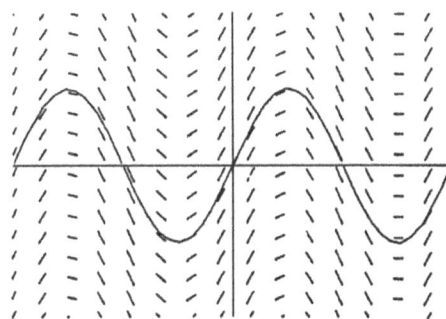

$[-2\pi,2\pi] \times [-2,2]$

(C)

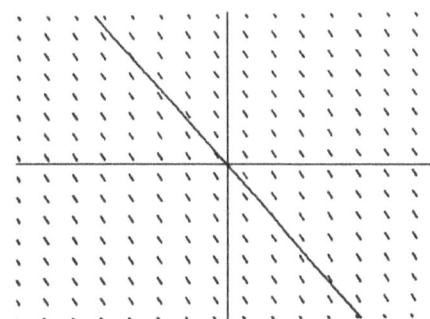

$[-2,2] \times [-2,2]$

(D)

Solution.

7.3   Approximating Solutions Using Euler's Method

So far, we've learnt how to sketch the slope field of a differential equation by taking the slopes of any single points. Then we can use these slope to make linear approximations by **Euler's Method**.

Theorem 7.3 (Euler's Method).

Euler's Method uses a linear approximation with increments, $h = \Delta x$, for solving differential equation with a given initial value x_0.

Define: (1) step size $\Delta x = x_n - x_{n-1}$ (2) slope $f'(x_n) = y_n{}'$ (3) initial value $(x_0, y_0) = (x_0, f(x_0))$

Then the Euler's Method indicates that:

First approximation	$y_1 \approx y_0 + \Delta x \cdot y_0{}'$
Second approximation	$y_2 \approx y_1 + \Delta x \cdot y_1{}'$
Third approximation	$y_3 \approx y_2 + \Delta x \cdot y_2{}'$

$$\vdots$$

$$y_{n+1} = y_n + \Delta x \cdot y_n{}'$$

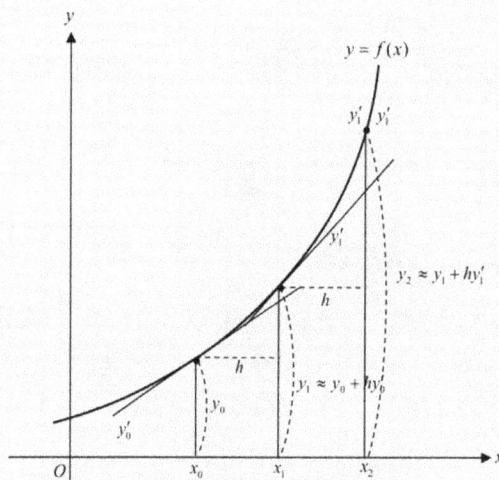

Example (1). Use Euler's method with a step size of $h = 1/4$ (or $\Delta x = 1/4$) to approximate $y(2)$ if $\dfrac{dy}{dx} = y + 1$ and point $(1,1)$ belongs to the graph of the solution of the differential equation.

Solution.

Example (2). Use Euler's method with step size $h = 0.2$ to approximate $y(1.6)$ if $\dfrac{dy}{dx} = \dfrac{x+y}{x}$ and $y(1) = 2$.

Solution.

7.4   Finding Solutions Using Separation of Variables

In this sections, we are going to introduce a solution technique for finding exact solutions to **separable differential equations**. These equations are common in a wide variety of disciplines, including physics, chemistry, and engineering.

Theorem 7.4. Separable Differential Equations A first order differential equation $y' = f(x, y)$ is called a separable equation if the function $f(x, y)$ can be factored into the product (or quotient) of two functions of x and y:

$$\frac{dy}{dx} = f(x) \cdot g(y) = \frac{f(x)}{h(y)}, \quad \text{where } g(y) = \frac{1}{h(y)}$$

These equations can be solve by **separating variables**.

Claim 7.5 (Separating Variable).

The separable differential equations $\dfrac{dy}{dx} = f(x) \cdot g(y)$ can be solve by the following steps:

1. Separate the variables: $\dfrac{1}{g(y)} \, dy = f(x) \, dx$.

2. Integrate both sides: $\displaystyle\int \frac{1}{g(y)} \, dy = \int f(x) \, dx$

3. Combine the constant: $H(y) = F(x) + C$.

4. Solve for y to get a general solution.

5. Substitute the given condition to get a particular solution.

6. Verify your result by differentiating.

Example (1). Given $\dfrac{dy}{dx} = 4x^3 y^2$ and $y(1) = -\dfrac{1}{2}$, solve the differential equation.

Solution.

Example (2). Find the general solution of the differential equation $\dfrac{\mathrm{d}y}{\mathrm{d}x} = \dfrac{2xy}{x^2+1}$.

Solution.

Example (3). Find the particular solution of the differential equation $\dfrac{\mathrm{d}y}{\mathrm{d}x} = -2y^2$, with initial condition $y(0) = -\dfrac{1}{2}$.

Solution.

7.5   Exponential Models with Differential Equations

In many contextual situations, we need to use the differential equations for modeling changes in variables. The most typical one is the **exponential growth and decay**.

Theorem 7.6 (Exponential Growth and Decay).

A quantity $y(t)$ is said to have an **exponential growth model** if it increases at a rate *proportional to the amount present*. It is said to have an **exponential decay model** if it decreases at a rate that is *proportional to the amount present*.

In the exponential growth model, y satisfies the differential equation:

$$\frac{\mathrm{d}y}{\mathrm{d}t} = ky$$

In the exponential decay model, y satisfies the differential equation:

$$\frac{\mathrm{d}y}{\mathrm{d}t} = -ky$$

where k is a positive constant.

Example (1). Show that the exponential model can be expressed as the differential equation $\dfrac{\mathrm{d}y}{\mathrm{d}t} = ky$, where $t \geq 0$, $y \geq 0$, and find the particular solution with initial condition $y(0) = y_0$

Solution.

Corollary 7.7 (Doubling Time and Half Life).

For an exponential growth modeled by $\dfrac{\mathrm{d}y}{\mathrm{d}t} = kt$ and $y = y_0 \cdot e^{kt}$, the doubling time is given by:

$$kt = \ln 2 \ \Rightarrow \ t = \frac{\ln 2}{k}$$

For an exponential decay modeled by $\dfrac{\mathrm{d}y}{\mathrm{d}t} = -kt$ and $y = y_0 \cdot e^{-kt}$, the half life is given by:

$$kt = \ln 2 \ \Rightarrow \ t = \frac{\ln 2}{k}$$

Example (2). The radioactive isotope Indium-111 is often used for diagnosis and imaging in nuclear medicine. Its half life is 2.8 days. What was the initial mass of the isotope before decay, if the mass in 2 weeks was 5 g.

Solution.

Corollary 7.8 (Initial Value and Growth/Decay constant).

For an exponential change modeled by $\dfrac{\mathrm{d}y}{\mathrm{d}x} = kt$ or $y = y_0 \cdot e^{kt}$, given that:

$$y_1 = y_0 \cdot e^{kt_1}$$
$$y_2 = y_0 \cdot e^{kt_2}$$

Then the Growth/Decay constant k and Initial Value y_0 can be derived by:

$$k = \frac{\ln y_2 - \ln y_1}{t_2 - t_1}, \qquad \ln y_0 = \frac{t_2 \ln y_1 - t_1 \ln y_2}{t_2 - t_1}$$

It can be solved by dividing the two equation.

Example (3). The rate of growth of population of flies is proportional to the size of population. In an experiment, it was observed that there were 200 flies after the second day and 1000 flies after the fourth day. How many flies were there in the original population?

Solution.

7.6   Logistic Models with Differential Equations

In the previous section, we have studied the exponential growth and decay, which is the simplest model. A more realistic model includes other factors that affect the growth. In this section, we are focusing on the logistic differential equation and see how it applies to various contexts.

Theorem 7.9 (Logistic Model).

In population models, the size of population approaches a positive constant M, called the **carrying capacity** of the system. One model with this property is provided by the logistic differential equation.

$$\frac{dP}{dt} = kP\left(1 - \frac{P}{M}\right) \quad \text{or} \quad \frac{dP}{dt} = \lambda P(M - P)$$

These two equations are equivalent since $k = m\lambda$.

The solution to the **logistic differential equation** is:

$$P = \frac{M}{1 + Ae^{-kt}}$$

where M is the carrying capacity, k is the growth constant and A is an arbitrary constant for a general solution.

Corollary 7.10 (Properties of Logistic Model).

The graph of the logistic growth is given by the following curve:

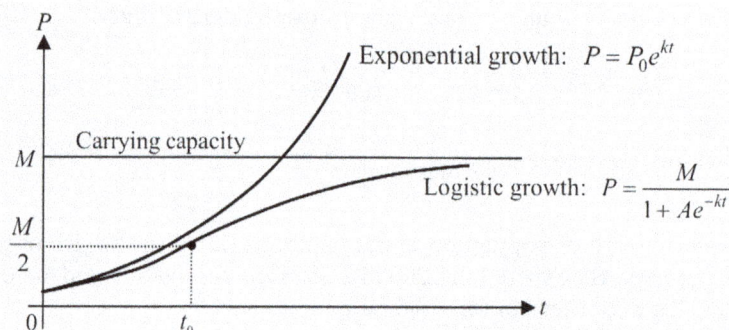

It has the following properties:

1. At time $t = 0$, the initial value of P is: $P_0 = \dfrac{M}{1 + A}$

2. At time $t_0 = \dfrac{\ln A}{k}$, the population is growing the fastest, where $P = \dfrac{M}{2}$

3. The point $\left(t_0, \dfrac{M}{2}\right)$ is the point of inflection.

4. $\lim\limits_{t \to \infty} P(t) = M$ and $\lim\limits_{t \to \infty} \dfrac{dP}{dt} = 0$

Example (1). If the population of a kind of cell is modelled by $\dfrac{dP}{dt} = 20P(400 - P)$, find the population when the population grows the fastest.

Solution.

Example (2). The growing of a population is modeled by the following function: (P increases according to the logistic model):
$$\frac{\mathrm{d}P}{\mathrm{d}t} = \frac{4}{5}P\left(1 - \frac{P}{20}\right)$$

(1) Find $P(t)$ if $P(0) = 5$.

(2) What is $\lim_{t \to \infty} P(t)$

(3) For what value of P is the population growing the fastest?

(4) For what value of t is the population growing the fastest?

Solution.

Example (3). A lake is stocked with 500 fish. If the population increases according to the logistic curve $y = \dfrac{10000}{1 + Ae^{-t/5}}$, where y is the fish population and t is measured in months.

(1) At what rate is the fish population changing at the end of the fifth month?

(2) After how many months is the population increasing the most rapidly?

Solution.

8 Applications of Integration

8.1 Finding the Average Value of a Function on an Interval

In UNIT 5, we have discussed about the **mean value theorem** (MVT), which guarantees a point $x = c$, such that $f'(c)$ is equal to the average rate of change. By the fundamental theorem of calculus, we can also deduce the mean value theorem for integrals.

Theorem 8.1 (Mean Value Theorem for Integrals).

If a function f is continuous on the interval $[a, b]$, then there exists at least one point c in $[a, b]$ such that

$$\int_a^b f(x) \, dx = f(c)(b - a)$$

Let f be a continuous non-negative on $[a, b]$, and let m and M be minimum and maximum values of $f(x)$ on this interval.

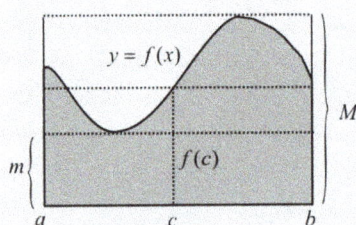

Since $(b - a)m < \int_a^b f(x) \, dx < (b - a)M$, there is a rectangle over the interval $[a, b]$ of some

appropriate height $f(c)$ between m and M whose area is $\int_a^b f(x) \, dx = f(c)(b - a)$.

The previous THEOREM can be modified to calculate the **average value** of a function.

Corollary 8.2 (The Average Value of a Function).

If a function f is continuous on the interval $[a, b]$, then, the average value of a function f over the interval $[a, b]$ is given by

$$f_{\text{average}} = \frac{1}{b - a} \int_a^b f(x) \, dx$$

Example (1). The graph of a function f is shown in figure. Find the average value of f on $[0, 4]$.

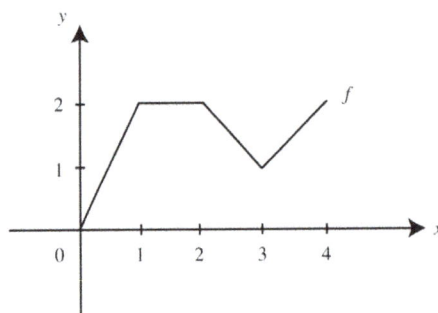

Solution.

Example (2). Let f be the function that is defined for all real number x. f has following properties.

$$\text{(i) } f''(x) = 24x - 18 \qquad \text{(ii) } f'(1) = -6 \qquad \text{(iii) } f(2) = 0$$

(a) Find each x such that the line tangent to the graph of f at $(x, f(x)$ is horizontal.

(b) Find $f(x)$.

(c) Find the average value of f on the interval $1 \leq x \leq 3$.

Solution.

Example (3). Let $f(x) = 10\pi x^2$ and $g(x) = k^2 \sin\left(\dfrac{\pi x}{2k}\right)$ for $k > 0$.

(a) What is the average value of f on $[1, 4]$?

(b) For what value of x will the average value of g on $[0, k]$ be equal to the average value of f on $[0, 4]$?

Solution.

8.2   Connecting Position, Velocity and Acceleration Using Integrals

In UNIT 4, we have learnt how to derive velocity and acceleration by differentiation. In this section, we are going to use integrals to reverse the process.

Theorem 8.3. For a particle moving on a straight line, the position $x(t)$, velocity $v(t)$ and acceleration $a(t)$ have the following relationships:

1. Position: $x(t) = \displaystyle\int v(t)\,dt$.

2. Velocity: $v(t) = \dfrac{dx}{dt}$ and $\Delta v = \displaystyle\int_{t_1}^{t_2} a(t)\,dt$

3. Acceleration: $a(t) = \dfrac{dv}{dt}$.

4. Velocity at t_2: $v(t_2) = \displaystyle\int_{t_1}^{t_2} a(t)\,dt$ Speed at t_2: $|v(t)|$

5. Displacement in $[t_1, t_2]$: $\displaystyle\int_{t_1}^{t_2} v(t)\,dt$ Total Distance in $[t_1, t_2]$: $\displaystyle\int_{t_1}^{t_2} |v(t)|\,dt$

Example (1). The graph of the velocity function of a moving particle is shown in the figure below. What is the total distance traveled by the particle during $0 \le t \le 12$?

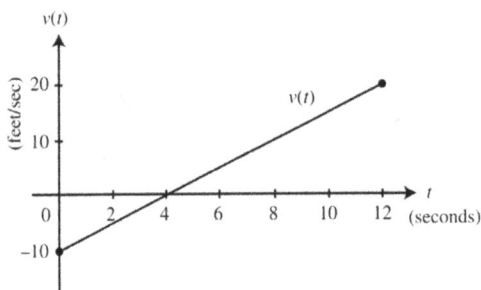

Solution.

Example (2). The velocity of a moving particle on a line is $v(t) = t^2 + 3t - 10$ for $0 \le t \le 6$. Find (a) the displacement during $0 \le t \le 6$, and (b) the total distance traveled during $0 \le t \le 6$.

Solution.

Example (3). The velocity function of a moving particle on a line is $v(t) = 3\cos(2t)$ for $0 \le t \le 2\pi$.

(a) Determine when the particle is moving to the right.

(b) Determine when the particle stops.

(c) The total distance traveled by the particle during $0 \le t \le 2\pi$.

You might use your calculator to solve this problem.

Solution.

8.3 Finding the Area Bounded by a Curve

As is stated in UNIT 6, the value of the definite integral represents the **signed area** bounded by the curve, then we can use the following techniques to solve area problems using integration.

> **Theorem 8.4** (Area Bounded by a Curve).
>
> If $y = f(x)$ is continuous and non-negative on $[a, b]$, then the area under the curve from a to b is:
>
> $$\text{Area} = \int_a^b f(x)\, dx$$
>
> If $y = f(x)$ is continuous and $f(x) < 0$ on $[a, b]$, then the area under the curve from a to b is:
>
> $$\text{Area} = -\int_a^b f(x)\, dx$$
>
> If $x = g(y)$ is continuous and non-negative on $[c, d]$, then the area bounded by the curve of g from c to d is:
>
> $$\text{Area} = \int_c^d g(y)\, dy$$

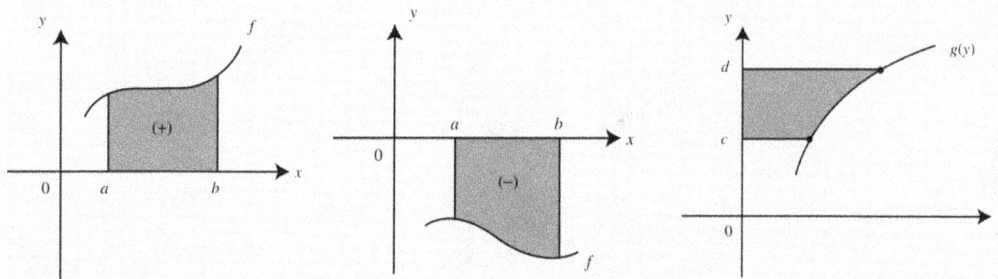

> **Example** (1). Find the area of the region bounded by the graph of $f(x) = x^2 - 1$, the lines $x = -2$ and $x = 2$, and the x-axis.
>
> **Solution.**

> **Example** (2). Find the area of the region bounded by $x = y^2$, $y = -1$, $y = 3$ and the y-axis.
>
> **Solution.**

Example (3). Let $F(x) = \displaystyle\int_0^x f(x)\, \mathrm{d}x$, where the graph of f is given in the figure below.

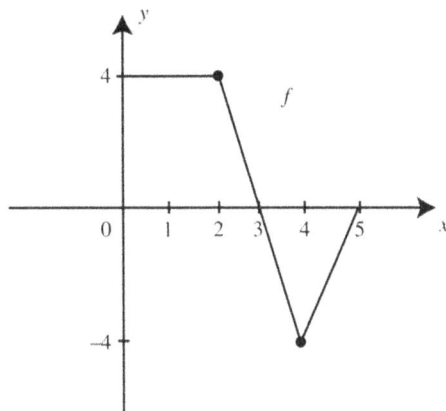

(1) Evaluate $F(0)$, $F(3)$ and $F(5)$.

(2) On what intervals is F increasing?

(3) At what value of x does F have the maximum value?

(4) On what intervals is F concave up?

Solution.

8.4   Finding the Area Between Curves

In this section, We would extend the notion of the area under a curve and consider the area of the region between two curves.

Theorem 8.5 (Area Between Two Curves).

The area between curves expressed as function of x is given by

$$A = \int_a^c [f(x) - g(x)] \, \mathrm{d}x + \int_c^d [g(x) - f(x)] \, \mathrm{d}x = \int_a^d (|\text{upper curve} - \text{lower curve}| \, \mathrm{d}x$$

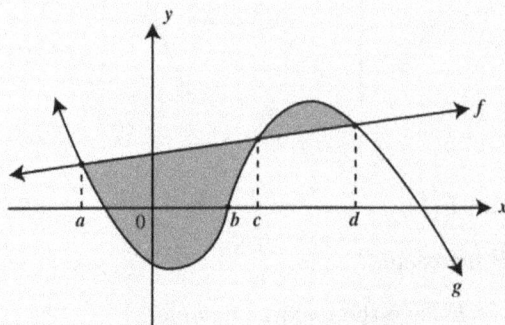

The area between curves expressed as function of y is given by

$$A = \int_m^n [f(y) - g(y)] \, \mathrm{d}y = \int_a^d |\text{right curve} - \text{left curve}| \, \mathrm{d}x$$

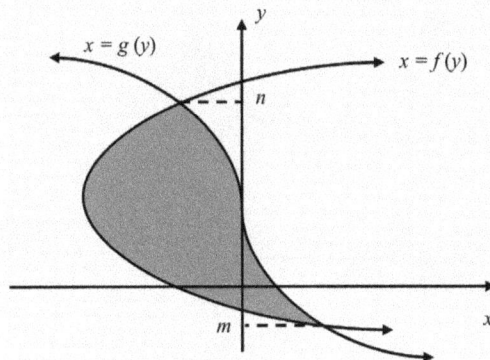

Example (1). Find the area of the regions bounded by the graphs of $f(x) = (x-1)^3$ and $g(x) = x-1$.

Solution.

Example (2). Find the area of the regions bounded by the graphs of $x = y^2 - 4y$ and $x = y$.

Solution.

8.5   Volume with Cross Sections

Besides the area, we can also calculate the volume of some solids form by the curves using integration. In this section, we will learn how to find the volume of a solid object that has known cross sections such as squares, rectangles, triangles and semicircles.

Theorem 8.6 (Volume of Solids with Known Cross Sections).

If A is the area of a cross section of a solid and A is continuous on $[a, b]$

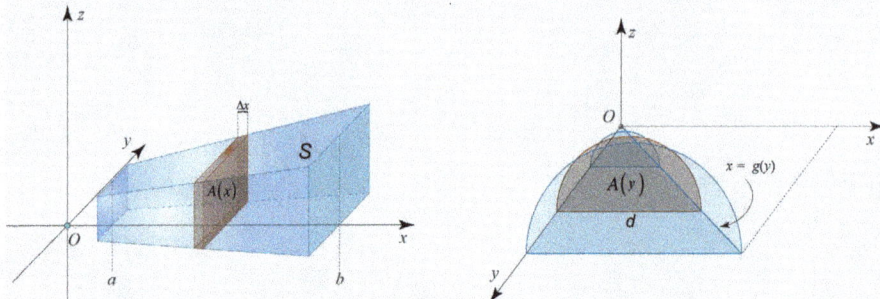

If the cross section is perpendicular to x-axis, then the volume of the solid from a to b is:

$$\int_a^b A(x) \, \mathrm{d}x$$

If the cross section is perpendicular to y-axis, then the volume of the solid from a to b is:

$$\int_a^b A(y) \, \mathrm{d}y$$

Example (1). The volume of the solid whose base is the region of the circle $x^2 + y^2 = 4$ whose cross sections taken perpendicular to the x-axis are squares.

Solution.

Example (2). Let R be the region in the first quadrant enclosed by $y = \sin x$, $y = \cos x$, and $x = 0$. Find the volume of the solid generated whose base is the region R and whose cross sections, perpendicular to the x-axis, are squares.

Solution.

Example (3). The base of a solid is the region enclosed by the ellipse $\dfrac{x^2}{4} + \dfrac{y^2}{25} = 1$. The cross sections are perpendicular to the x-axis and are isosceles right triangles whose hypotenuses are on the ellipse. Find the volume of the solid.

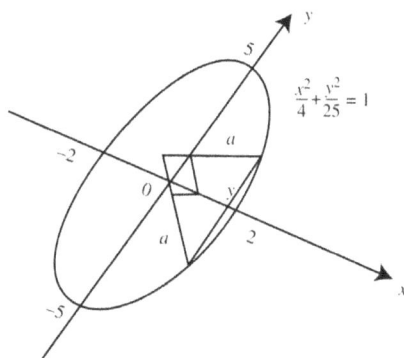

Solution.

Example (4). The base of a solid is the region enclosed by a triangle whose vertices are $(0,0)$, $(4,0)$, and $(0,2)$. The cross sections are semicircles perpendicular to the x-axis. Using a calculator, find the volume of the solid.

Solution.

8.6   Volume with Disc Method

If a region in the plane is revolved about a line in the same plane, the resulting object is known as a solid of revolution. For solid shapes, we may use the disc method to calculate the volume.

Theorem 8.7 (The Disc Method).

The **disc method** is based on slicing the solid into pieces with cross section plane perpendicular to the coordinate axis.

By this method, the volume of each disc is given as follows

$$dV = \pi r^2 \, dx \quad \text{or} \quad dV = \pi r^2 \, dy$$

By integrating both sides, we can calculate the total volume.

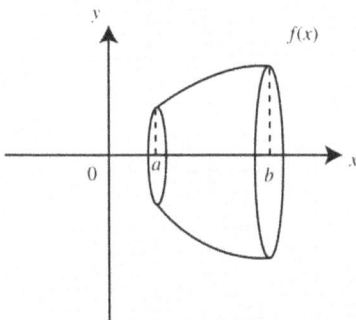

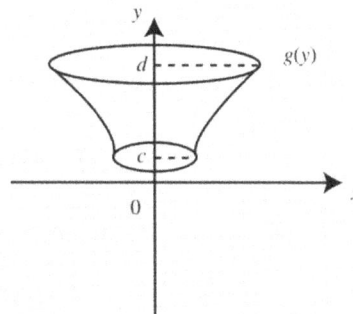

If the solid is formed by revolving about x-axis:

$$V = \pi \int_a^b (f(x))^2 \, dx, \quad \text{where } f(x) = \text{radius.}$$

If the solid is formed by revolving about y-axis:

$$V = \pi \int_c^d (g(y))^2 \, dy, \quad \text{where } g(y) = \text{radius.}$$

Example (1). Let R be the region enclosed by $y = \tan x$, the x-axis, and $x = \dfrac{\pi}{3}$. Find the volume of the solid formed by revolving the region bounded by the graphs about the x-axis.

Solution.

Example (2). Find the volume of the solid generated by revolving about the y-axis the region in the first quadrant bounded by the graph of $y = x^2$, the y-axis, and the line $y = 6$.

Solution.

Theorem 8.8 (The Disc Method Con't).

The **disc method** can also be used to deduce the volume of revolution about the line parallel to the coordinate axis.

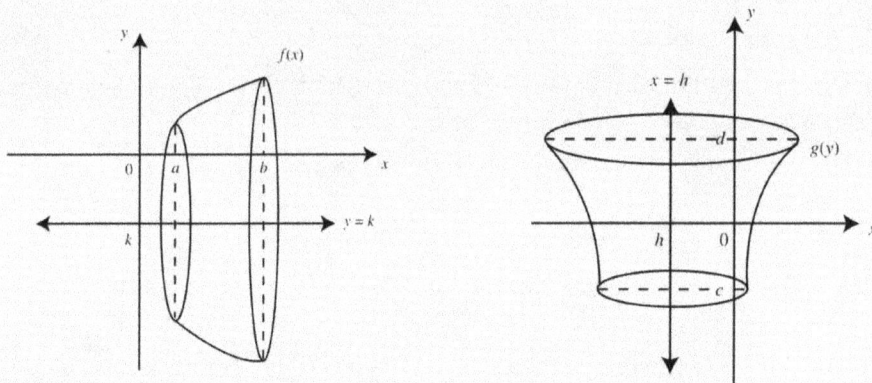

If the solid is formed by revolving about $y = k$:

$$V = \pi \int_a^b (f(x) - k)^2 \, \mathrm{d}x, \quad \text{where } |f(x) - k| = \text{radius.}$$

If the solid is formed by revolving about $x = h$:

$$V = \pi \int_c^d (g(y) - h)^2 \, \mathrm{d}y, \quad \text{where } |g(y) - h| = \text{radius.}$$

Example (3). Find the volume of the solid generated when the region under the curve $y = x^2$ over the interval $[0, 2]$ is rotated about the line $x = 2$.

Solution.

Example (4). Using a calculator, find the volume of the solid generated by revolving about the line $y = -3$, the region bounded by the graph of $y = e^x$, the y-axis, and the lines $x = \ln 2$ and $y = -3$.

Solution.

8.7 Volume with Washer Method

The disk method can be extended to find the volume of a hollow solid of revolution, which is usually called the **washer method**.

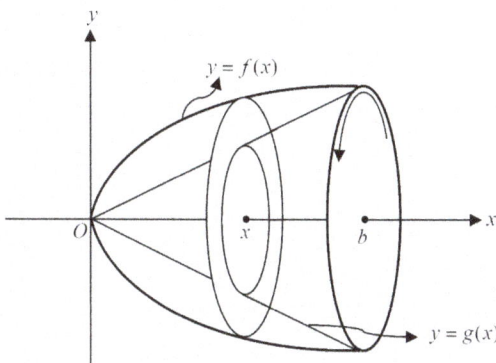

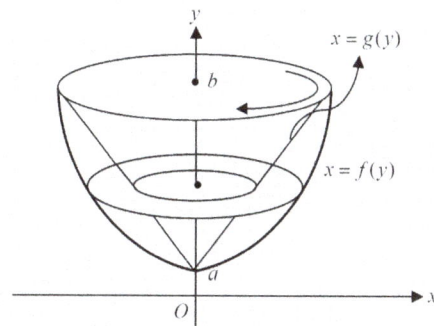

> **Theorem 8.9** (Washer Method).
>
> Assuming that the functions $f(x)$ and $g(x)$ are continuous on the interval $[a, b]$ and consider a region that is bounded by two curves $y = f(x)$ and $y = g(x)$ between $x = a$ and $x = b$.
>
> The area of each cross section is:
>
> $$dV = A(x) = \pi \left(R^2 - r^2 \right) = \pi \left((f(x))^2 - (g(x))^2 \right)$$
>
> Therefore, the volume of the solid formed by revolving the region about the x-axis is
>
> $$V = \pi \int_a^b \left[(f(x))^2 - (g(x))^2 \right] \, dx$$
>
>
>
> Assuming that the functions $f(y)$ and $g(yx)$ are continuous on the interval $[a, b]$ and consider a region that is bounded by two curves $x = f(y)$ and $x = g(y)$ between $y = a$ and $y = b$.
>
> The area of each cross section is:
>
> $$dV = A(x) = \pi \left(R^2 - r^2 \right) = \pi \left((f(x))^2 - (g(x))^2 \right)$$
>
> Therefore, the volume of the solid formed by revolving the region about the y-axis is
>
> $$V = \pi \int_a^b \left[(f(y))^2 - (g(y))^2 \right] \, dy$$

If the axis of revolution is a line parallel to the coordinate axis, then we may also deduce the volume with **washer method**.

> **Corollary 8.10** (Washer Method with Other Axis of Revolution).
>
> The volume of a solid (with a hole in the middle) generated by revolving a region bounded by 2 curves is given by the following:
>
> 1. If the axis of revolution is $y = h$, then:
>
> $$V = \pi \int_a^b \left[(f(x) - h)^2 - (g(x) - h)^2 \right] \, dx$$
>
> 2. If the axis of revolution is $x = k$, then:
>
> $$V = \pi \int_a^b \left[(f(y) - k)^2 - (g(y) - k)^2 \right] \, dy$$

Example (1). Let R be the region in the first quadrant enclosed by the graph of $y = \sqrt{6x + 4}$, the line $y = 2x$, and the y-axis. Find the volume of the solid generated when R is revolved about the x-axis.

Solution.

Example (2). Using the Washer Method, find the volume of the solid generated by revolving the region bounded by $y = x^2$ and $x = y^2$ about the y-axis.

Solution.

8.8   The Arc Length of a Smooth, Planer Curve

In this section we are going to focus on computing the arc length of a function by definite integral, which requires us to split the curved arc into many short segments.

Theorem 8.11 (Arc Length of a Curve).

If the function $y = f(x)$ is differentiable on the interval $[a, b]$, then the length of the curve between $x = a$ and $x = b$ is

$$L = \int_a^b \sqrt{1 + \left(\frac{dy}{dx}\right)^2}\ dx = \int_a^b \sqrt{1 + (f'(x))^2}\ dx$$

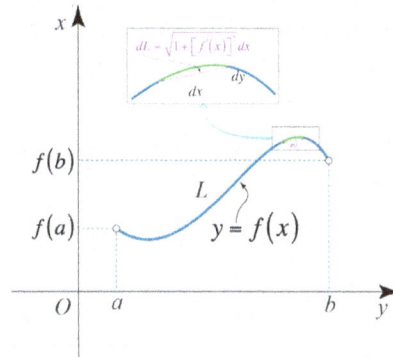

If the function $x = f(y)$ is differentiable on the interval $[a, b]$, then the length of the curve between $y = a$ and $y = b$ is

$$L = \int_a^b \sqrt{1 + \left(\frac{dx}{dy}\right)^2}\ dy = \int_a^b \sqrt{1 + (f'(y))^2}\ dy$$

Example (1). Find the arc length of the function $y = \dfrac{1}{3}\left(x^2 + 2\right)^{\frac{3}{2}}$ over the interval $[0, 3]$.

Solution.

Example (2). Find the arc length of the function $x = \dfrac{1}{3}\sqrt{y}(y - 3)$ from $y = 1$ to $y = 9$.

Solution.

9   Parametric Equations, Polar Coordinates and Vector-Valued Functions

9.1   Defining and Differentiating Parametric Equations

In this section, we are going to find out the techniques of differentiating parametric equations and discuss about the practical meanings of the derivatives.

Theorem 9.1 (Derivatives of Parametric Equations).

For a parametric equations $x = f(t)$, $y = g(t)$, the first derivative is given by

$$\frac{dy}{dx} = \frac{dy/dt}{dx/dt}, \text{ where } \frac{dx}{dt} \neq 0$$

1. A horizontal tangent line occurs where $\dfrac{dy}{dt} = 0$ and $\dfrac{dx}{dt} \neq 0$.

2. A vertical tangent line occurs where $\dfrac{dx}{dt} = 0$ and $\dfrac{dy}{dt} \neq 0$.

The **second derivative** of parametric equations is given by

$$\frac{d^2y}{dx^2} = \frac{dy'}{dx} = \frac{dy'/dt}{dx/dt} = \frac{d\left(\frac{dy}{dx}\right)/dt}{dx/dt}$$

Example (1). For parametric equations $x = t^2$ and $y = t^3 - 3t$ $(t \geq 0)$

(a) Find the point where horizontal tangent occurs.

(b) Find the point where vertical tangent occurs.

Solution.

Example (2). If parametric equations $x = \sec t$ and $y = \tan t$ on $\left[-\dfrac{\pi}{2}, \dfrac{\pi}{2}\right]$, find the equation of tangent line at $t = \dfrac{\pi}{4}$.

Solution.

Example (3). Find $\dfrac{\mathrm{d}y}{\mathrm{d}x}$ and $\dfrac{\mathrm{d}^2 y}{\mathrm{d}x^2}$ at point $(1,1)$ on the curve given by the parametric equations $x = t^2$, $y = t^3$.

Solution.

Example (4). If $x = \ln(2t)$, $y = \ln(3t)^4$, find $\dfrac{\mathrm{d}^2 y}{\mathrm{d}x^2}$ in terms of t

Solution.

9.2   Finding Arc Lengths of Parametric Curves

In UNIT 8, we have learnt how to calculate the arc length for a curve defined by a function. The arc length of a parametric curve can also be determined in a similar way.

Theorem 9.2 (Arc Length of Parametric Curves).

If a smooth curve C is given by $x = f(t)$ and $y = g(t)$, then the arc length over the interval $a \leq t \leq b$ is given by:

$$L = \int_a^b \sqrt{\left(\frac{\mathrm{d}x}{\mathrm{d}t}\right)^2 + \left(\frac{\mathrm{d}y}{\mathrm{d}t}\right)^2}\, \mathrm{d}t$$

Example (1). Find arc length of the function over the indicated interval.

(a) $x = \dfrac{1}{3}t^3$, $y = \dfrac{1}{2}t^2$, $0 \leq t \leq 1$

(b) $x = e^{-t}\cos t$, $y = e^{-t}\sin t$, $0 \leq t \leq \dfrac{\pi}{2}$

Solution.

9.3   Vector-Valued Functions

In the previous UNITS, we have learnt how to apply the differentiation and integration to solve the motion problems in a straight line. For multidimensional motion (2D and 3D), we may use a vector to represent the position, velocity and even acceleration of a moving particle.

In this section, we are going to discuss about the 2-dimensional vectors and their properties in algebra and calculus.

> **Lemma 9.3** (Vector and Vector Arithmetic).
>
> In two dimensions, a vector is usually described algebraically by two parameters in the form $\langle a, b \rangle$. They have the following properties
>
> 1. Magnitude of Vector:
>
>   The **magnitude** or **modulus** of a vector $\langle a, b \rangle$ is given by a non-negative real number
>
> $$|\langle a, b \rangle| = \sqrt{a^2 + b^2}$$
>
> 2. Vector Addition and Scalar Multiplication
>
>   Let $\mathbf{u} = \langle u_1, u_2 \rangle$, and $\mathbf{v} = \langle v_1, v_2 \rangle$, and k be a real number(scalar).
>
>   The **sum** (or **resultant**) of the vectors $\mathbf{u}$ and v is the vector:
>
> $$\mathbf{u} + \mathbf{v} = \langle u_1 + u_2, v_1 + v_2 \rangle$$
>
>   The **product** of the scalar k and the vector $\mathbf{u}$ is
>
> $$k\mathbf{u} = k\langle u_1, u_2 \rangle = \langle ku_1, ku_2 \rangle$$
>
>   The **opposite vector** of $\mathbf{v}$ is $-\mathbf{v} = (-1)\mathbf{v}$. We define vector subtraction by
>
> $$\mathbf{u} - \mathbf{v} = \mathbf{u} + (-\mathbf{v}).$$

By using the vector notation, we can now solve the 2D motion problems via differentiating and integrating **vector-valued functions**.

> **Theorem 9.4** (Differentiating Vector-Valued Functions).
>
> Suppose a particle moves along a smooth curve in the plane so that its position at any time t is $(x(t), y(t))$, where $x(t)$ and $y(t)$ are differentiable functions of t.
>
> 1) The particle's **position vector** is $\mathbf{r}(t) = \langle x(t), y(t) \rangle$.
>
> 2) The particle's **velocity vector** is $\mathbf{v}(t) = \left\langle \dfrac{\mathrm{d}x}{\mathrm{d}t}, \dfrac{\mathrm{d}y}{\mathrm{d}t} \right\rangle$.
>
> 3) The particle's **speed** is $|\mathbf{v}(t)|$. It is a scalar.
>
> 4) The particle's **acceleration vector** is $\mathbf{a}(t) = \left\langle \dfrac{\mathrm{d}^2 x}{\mathrm{d}t^2}, \dfrac{\mathrm{d}^2 y}{\mathrm{d}t^2} \right\rangle$.
>
> 5) The particle's **direction of motion** is usually described by the **unit direction vector** $\dfrac{\mathbf{v}}{|\mathbf{v}|}$.

> **Claim 9.5.**
>
> In motion problems, **displacement** and **velocity** are described as *vectors*, but **speed** and **distance** are described as *scalars*.

Example (1). A particle moves in an elliptical path so that its position at any time $t \geq 0$ is given by $\langle 4\sin t, 2\cos t \rangle$.

(a) Find the velocity and acceleration vectors.

(b) Find the velocity, acceleration, speed, and direction of motion at $t = \dfrac{\pi}{4}$.

Solution.

Theorem 9.6 (Integrating Vector-Valued Functions).

Suppose a particle moves along a path in the plane so that its velocity at any time t is $\mathbf{v}(t) = \langle v_1(t), v_2(t) \rangle$, where v_1 and v_2 are integrable functions of t.

The displacement from $t = a$ to $t = b$ is given by the vector

$$\left\langle \int_a^b v_1(t)\, dt, \int_a^b v_2(t)\, dt \right\rangle.$$

The preceding vector is added to the position at time $t = a$ to get the position at time $t = b$.

The distance traveled from $t = a$ to $t = b$ is

$$\int_a^b |\mathbf{v}(t)|\, dt = \int_a^b \sqrt{(v_1(t))^2 + (v_2(t))^2}\, dt.$$

Which is the same as the arc length of the curve from $t = a$ to $t = b$.

Example (2). A particle moves in the plane with velocity vector $\mathbf{v}(t) = \langle t - 3\pi\cos(\pi t), 2t - \pi\sin(\pi t) \rangle$. At $t = 0$, the particle is at the point $(1, 5)$. Use your calculator, find:

(a) The position of the particle at $t = 4$.

(b) The total distance traveled by the particle from $t = 0$ to $t = 4$?

Solution.

Example (3). For time $t \geq 0$, a particle moves in the xy-plane with position $\langle (x(t), y(t) \rangle$ and velocity vector $\left\langle (t-1)e^{t^2}, \sin t^{1.25} \right\rangle$. At time $t = 0$, the position of the particle is $(-2, 5)$. Use your calculator, find:

(a) The speed of the particle at time $t = 1.2$ and the acceleration vector of the particle at time $t = 1.2$.

(b) The total distance traveled by the particle over the time interval $0 \leq t \leq 1.2$.

(c) Find the coordinates of the point at which the particle is farthest to the left for $t \geq 0$. Explain why there is no point at which the particle is farthest to the right for $t \geq 0$.

Solution.

9.4   Polar Coordinates

Apart from the Cartesian coordinate system that uses the horizontal and vertical coordinates x and y to represent points, the polar coordinates specifies the location of a point by its distance and the angle made between the line segment and the positive x-axis.

Lemma 9.7 (Polar Coordinate).

The **polar coordinate** $P\,(r,\theta)$ defines a point P located at a distance of r from the origin (**pole**) and forms an angle θ between the line segment OP and the **polar axis**.

The relationship between the polar coordinate (r,θ) and the Cartesian coordinate (x,y) is as follows.

$$x = r\cos\theta,\ y = r\sin\theta \qquad r^2 = x^2 + y^2,\ \tan\theta = \frac{y}{x}$$

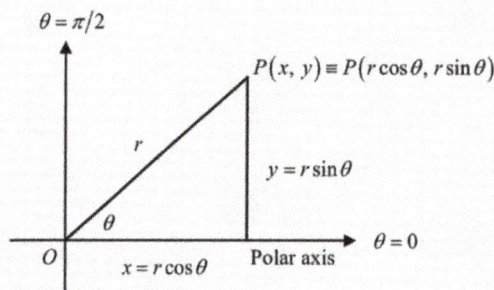

The **pole** is the *origin*, and the **polar axis** is positive x-axis.

Claim 9.8.

The r in the polar coordinate can be negative. And the same point might have many different representations in polar coordinates.

e.g. For $P(-5, \dfrac{\pi}{3}) = P(5, \dfrac{4\pi}{3}) = P(-5, \dfrac{7\pi}{3}) = \cdots$

Theorem 9.9 (Differentiation of Polar Curves).

If a polar curve is defined as a function $r = f(\theta)$, then the slope of the line tangent to the graph of $r = f(\theta)$ at the point (r, θ) is

$$\frac{\mathrm{d}y}{\mathrm{d}x} = \frac{\mathrm{d}y/\mathrm{d}\theta}{\mathrm{d}x/\mathrm{d}\theta} = \frac{f'(\theta)\sin\theta + f(\theta)\cos\theta}{f'(\theta)\cos\theta - f(\theta)\sin\theta}$$

Example (1). Find $\dfrac{\mathrm{d}y}{\mathrm{d}x}$ and slope of tangent at $\theta = \dfrac{\pi}{6}$ for $r = 1 + \cos\theta$

Solution.

Theorem 9.10 (Integration of Polar Curves).

The area of the region between the origin and the curve $r = f(\theta)$ for $\alpha \leq \theta \leq \beta$ is

$$A = \frac{1}{2} \int_{\alpha}^{\beta} r^2 \, \mathrm{d}\theta = \frac{1}{2} \int_{\alpha}^{\beta} (f(\theta))^2 \, \mathrm{d}\theta.$$

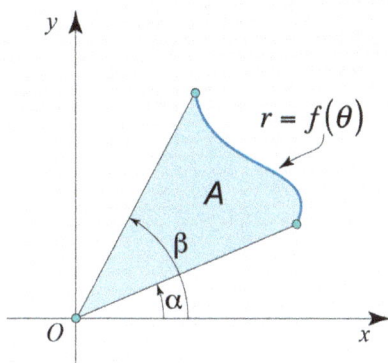

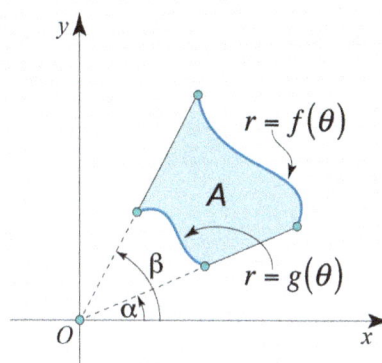

The area of the region between $r = f(\theta)$ and $r = g(\theta)$ for $\alpha \leq \theta \leq \beta$ is

$$A = \frac{1}{2} \int_{\alpha}^{\beta} R^2 - r^2 \, \mathrm{d}\theta = \frac{1}{2} \int_{\alpha}^{\beta} (f(\theta))^2 - (g(\theta))^2 \, \mathrm{d}\theta.$$

Example (2). The polar curve r is given by $r(\theta) = 1 - \sin\theta$ for $0 \leq \theta \leq 2\pi$. Find the area in the second quadrant enclosed by the coordinate axes and the graph of r.

Solution.

Example (3). Find the area of the region in the plane enclosed by the cardioid $r = 2(1 + \cos\theta)$.

Solution.

Example (4). Find the area inside the smaller loop of the limacon $r = 2\cos\theta + 1$.

Solution.

Example (5). Let R be the region inside the graph of the polar curve $r = 2$ and outside the graph of the polar curve $r = 2(1 - \sin\theta)$.

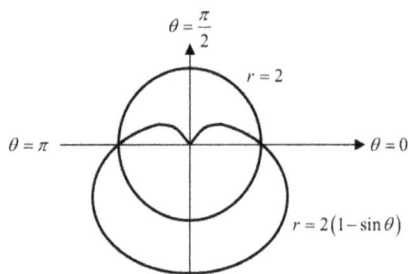

Solution.

Example (6).

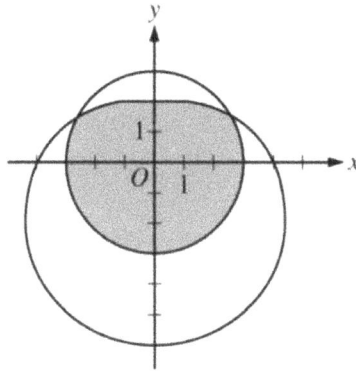

The graphs of the polar curves $r = 3$ and $r = 4 - 2\sin\theta$ are shown in the figure above. The curves intersect when $\theta = \dfrac{\pi}{6}$ and $\theta = \dfrac{5\pi}{6}$.

(a) Let S be the shaded region that is inside the graph of $r = 3$ and also inside the graph of $r = 4 - 2\sin\theta$. Find the area of S.

(b) A particle moves along the polar curve $r = 4 - 2\sin\theta$ so that at time t seconds, $\theta = t^2$. Find the time t in the interval $1 \le t \le 2$ for which the x-coordinate of the particle's position is -1.

(c) For the particle described in part (b), find the position vector in terms of t. Find the velocity vector at time $t = 1.5$.

Solution.

137

10   Infinite Sequences and Series

10.1   Defining Convergent and Divergent Infinite Sequence

In this UNIT, we are going to focus on the **sequences** and **series**. A sequence is nothing more than a list of numbers written in a specific order. The list may or may not have an infinite number of terms in it although we will be dealing exclusively with infinite sequences in this class.

> **Theorem 10.1** (Infinite Sequence).
>
> A sequence of real numbers is a function $f(n)$, whose domain is the set of positive integers. The values taken by the function are called the *terms* of the sequence.
>
> The set of values $a_n = f(n)$ is denoted by $\{a_n\}$. A sequence $\{a_n\}$ has the limit L if: $\lim_{n\to\infty} c_n = L$.
>
> If the *limit exists* and L is *finite*, we say that the sequence **converges**. Otherwise, the sequence **diverges**.

> **Example** (1). Write the formula for the following sequences and determine their limits (if exist).
>
> $$\frac{1}{3}, \frac{2}{4}, \frac{3}{5}, \frac{4}{6}, \frac{5}{7}, \cdots$$
>
> **Solution.**

> **Corollary 10.2** (Squeezing Theorem).
>
> Suppose $\lim_{n\to\infty} a_n = \lim_{n\to\infty} b_n = L$ and $\{c_n\}$ is a sequence such that $a_n \le c_n \le b_n$ for all n, then
>
> $$\lim_{n\to\infty} c_n = L$$
>
> The sequence $\{a_n\}$ is bounded if there is a number $M > 0$ such that for every positive n.
>
> *Every convergent sequence is bounded. Every unbounded sequence is divergent.*

> **Example** (2). Write the formula for the following sequences and determine their limits (if exist).
>
> $$1, -\frac{2}{2}, \frac{3}{4}, -\frac{4}{8}, \frac{5}{16}, \cdots$$
>
> **Solution.**

Example (3). Does the sequence $\left\{\dfrac{2n+3}{5n-7}\right\}$ converge or diverge?

Solution.

Example (4). Determine whether the sequence $\left\{\sqrt{n+2}-\sqrt{n+1}\right\}$ converges or diverges.

Solution.

Example (5). Does the sequence $\left\{\dfrac{n^2}{2^n}\right\}$ converge or diverge?

Solution.

10.2   Defining Convergent and Divergent Infinite Series

The infinite series is usually considered as the sum of infinitely many numbers in a sequence. To determine whether it converges we need to check the properties of them.

Theorem 10.3 (Infinite Series and Partial Sum).

Let $\{a_n\}$ be a sequence. Then the infinite sum:

$$\sum_{n=1}^{\infty} a_n = a_1 + a_2 + \cdots + a_n + \cdots$$

is called an **infinite series**, or, simply, series. The **partial sums** of the series are given by:

$$S_n = \sum_{k=1}^{n} a_k = a_1 + a_2 + \cdots + a_n$$

Where S_n is called the n-th partial sums of the series. If the partial sums $\{S_n\}$ converge to L as n approaches infinity, then we say that then we say that the infinite series converges to L, namely:

$$\sum_{n=1}^{\infty} a_n = L, \text{ if } \lim_{n \to \infty} S_n = L.$$

Otherwise, we say that the series $\sum_{n=1}^{\infty} a_n$ diverges.

Claim 10.4.

** Please do not confuse the convergence of a sequence and the convergence of a series.

** A *sequence* is convergent if $\lim_{n \to \infty} a_n = L$, a *series* is convergent when $\sum_{n=1}^{\infty} a_n = L$.

Example (6). Determine whether the series $\sum_{n=1}^{\infty} n$ converges or diverges.

Solution.

Example (7). Determine whether the series $\sum_{n=1}^{\infty} (-1)^{n+1} \cdot 2 = 2 - 2 + 2 \cdots$ converges or diverges.

Solution.

10.3 Telescoping Series, Geometric Series and Harmonic Series

In this section, we are going to focus on some *common series* and try to figure out the special properties for them.

Theorem 10.5 (Telescoping Series).

The series in the form

$$\sum_{n=1}^{\infty} \frac{1}{n(n+a)} = \sum_{n=1}^{\infty} \frac{1}{a}\left(\frac{1}{n} - \frac{1}{n+a}\right)$$

is called the **telescoping series**.

Example (1). Determine whether the following series converges or diverges and find the value that the series converge to:

(a) $\displaystyle\sum_{n=1}^{\infty} \frac{1}{n(n+1)}$.

(b) $\displaystyle\sum_{n=4}^{\infty} \frac{1}{n^2 - 4n + 3}$.

Solution.

Lemma 10.6 (Geometric Sequence).

A sequence of numbers $\{an\}$ is called a **geometric sequence** if the quotient of successive terms is a constant, called the **common ratio**. Namely:

$$\frac{a_{n+1}}{a_n} = r, \text{ for any integer } n$$

It is usually supposed that $r \neq 0$ and $r \neq 1$.

The general term for any geometric sequence is: $a_n = a_1 r^{n-1}$.

Example (2). For an arbitrary geometric sequence $\{a_n\}$ with first term a, and common ratio r ($r \neq 0$, 1), find the partial sum S_n for the series.

Solution.

Theorem 10.7 (Geometric Series).

A geometric series $\sum\limits_{n=1}^{\infty} ar^{n-1} = a + ar + ar^2 + ...ar^{n-1} + ... (a \neq 0)$ is also equivalent to the limit of the partial sum $S_n = \dfrac{a(1-r^n)}{1-r}$.

If and only if $|r| < 1$, $\lim\limits_{n\to\infty} r^n = 0$. The partial sum has limit:

$$\lim_{n\to\infty} S_n = \lim_{n\to\infty} \frac{a(1-r^n)}{1-r} = \frac{a}{1-r}.$$

Hence, we may say a geometric series **converges to** $\dfrac{a}{1-r}$ if and only if $|r| < 1$, where r is the common ratio of the geometric series.

Claim 10.8 (Convergence Test for Geometric Series).

For any geometric series, we only need to check the common ratio r to determine its convergence.

If $|r| < 1$, the geometric series converges, otherwise, the geometric series diverges.

Example (3). Find the sum of the series $\sum_{n=1}^{\infty} a_n = 1 - \dfrac{1}{\sqrt{2}} + \dfrac{1}{2} - \dfrac{1}{2\sqrt{2}} + \dfrac{1}{4} - \dfrac{1}{4\sqrt{2}} + ...$

Solution.

Example (4). Consider a recurring decimal $0.1313131313 \cdots$

(a) State the decimal using a geometric series.

(b) Find the exact fractional value of the decimal using convergence.

Solution.

Example (5). Find the range of x such that $\sum_{n=0}^{\infty} x^n$ converges, and evaluate the value of the series.

Solution.

Theorem 10.9 (Harmonic Series).

A series in the form of $\displaystyle\sum_{n=1}^{\infty} \frac{1}{n} = 1 + \frac{1}{2} + \frac{1}{3} + \frac{1}{4} + \cdots \frac{1}{n} + \cdots$ is called the **harmonic series**. Note that:

** The harmonic series *diverges*.

** The harmonic sequence *converges*.

Example (6). Show that the harmonic series diverges.

Solution.

10.4   Convergence Test for Positive Series

Starting from this section, we are going to discuss about the **convergence test** for infinite series. Before we start, let's introduce some important LEMMAS in determining convergent series.

Lemma 10.10 (Properties for the Convergence).

Given that a series $\sum a_n$ converges if and only if $\sum a_n$ is a finite value, then:

1. If $\displaystyle\sum_{n=1}^{\infty} a_n$, then $\displaystyle\lim_{n\to\infty} a_n = 0$.

2. A finite number of terms may be added to or deleted from a series without affecting its convergence, namely:

$$\sum_{n=1}^{\infty} a_n \text{ and } \sum_{n=m}^{\infty} a_n$$

   are both convergent or divergent, given that m is an arbitrary non-negative integer.

3. The series $\displaystyle\sum_{n=1}^{\infty} a_n$ and $\displaystyle\sum_{n=1}^{\infty} ka_n$, $(k \neq 0)$ both converges or diverges

4. If $\displaystyle\sum_{n=1}^{\infty} a_n$ and $\displaystyle\sum_{n=1}^{\infty} b_n$ converge, then $\displaystyle\sum_{n=1}^{\infty} (a_b + b_n$ also converges.

By the previous LEMMAS, we may start from the simplest test: the nth term test for divergence.

Theorem 10.11 (The nth Term Test).

If $\displaystyle\lim_{n\to\infty} a_n \neq 0$, then $\displaystyle\sum_{n=1}^{\infty} a_n$ diverges.

The contrapositive for this proposition is just the same as the first one in the previous LEMMA. If $\displaystyle\sum_{n=1}^{\infty} a_n$, then $\displaystyle\lim_{n\to\infty} a_n = 0$.

Example (1). Verify whether the series diverge or not.

(a) $\displaystyle\sum_{n=1}^{\infty} \frac{n}{2n+3}$

(b) $\displaystyle\sum_{n=1}^{\infty} \frac{e^n}{n^2}$

(c) $\displaystyle\sum_{n=1}^{\infty} \frac{1}{\cos\left(\frac{1}{n}\right)}$

Solution.

In the previous sections, we have learnt that the **harmonic series** is divergent. The p-series test is an extension of the idea of harmonic series test to determine the convergence or diverges of a infinite series defined as a *power function*.

Theorem 10.12 (Convergence Test for p-Series).

A series in the form of $\displaystyle\sum_{n=1}^{\infty} \frac{1}{n^p} = 1 + \frac{1}{2^p} + \frac{1}{3^p} + \frac{1}{4^p} + \cdots + \frac{1}{n^p} + \cdots$ is called the p-**series**.

For any arbitrary p-series $\displaystyle\sum_{n=1}^{\infty} \frac{1}{n^p}$ $\begin{cases} \text{If} & p > 1, \text{ then the series converge.} \\ \text{If} & p \leq 1, \text{ then the series diverge.} \end{cases}$

When $p = 1$, the series is *harmonic series*.

** *The **integral test** can be used to proof the validity of the test, which will be discussed in later.*

Example (2). Verify whether the series diverge or not.

(a) $\displaystyle\sum_{n=1}^{\infty} \frac{1}{n^e}$
(b) $\displaystyle\sum_{n=1}^{\infty} \frac{2}{n\sqrt{n}}$
(c) $\displaystyle\sum_{n=1}^{\infty} n^{\frac{1}{3}}$
(d) $\displaystyle\sum_{n=1}^{\infty} \frac{5}{\sqrt{n}}$

Solution.

There would be many series that is *close to* or *similar to* a p-series or geometric series. Here we can use the comparison test to determine the convergence.

Theorem 10.13 (Comparison Test).

For some series $\sum_{n=1}^{\infty} a_n$, we may need to construct a common series $\sum_{n=1}^{\infty} b_n$ (i.e. geometric series, p-series) that is similar to to make comparisons. The comparison test indicates:

If $0 < a_n \leq b_n$ for all n, then: $\sum_{n=1}^{\infty} b_n$ converge implies $\sum_{n=1}^{\infty} a_n$ converge.

If $0 < b_n \leq a_n$ for all n, then: $\sum_{n=1}^{\infty} b_n$ diverge implies $\sum_{n=1}^{\infty} a_n$ diverge.

Example (3). Determine whether the series $\sum_{n=1}^{\infty} \dfrac{n^2 - 1}{n^4}$ converges or diverges.

Solution.

Example (4). Determine whether the series $\sum_{n=1}^{\infty} \dfrac{n^2}{n^3 - 3}$ converges or diverges.

Solution.

Example (5). Determine whether the series $\displaystyle\sum_{n=1}^{\infty} \frac{e^{\frac{1}{n}}}{n^2}$ converges or diverges.

Solution.

Example (6). Determine whether the series $\displaystyle\sum_{n=1}^{\infty} \frac{2 + \cos n}{n^3}$ converges or diverges.

Solution.

Example (7). Determine whether the series $\displaystyle\sum_{n=0}^{\infty} \frac{2^n \sin^2(5n)}{4^n + \cos^2(3n)}$ converges or diverges.

Solution.

In some cases, we may fail to apply the **comparison test** despite we already find the similar common series to compare with. Then we may try the **limit comparison test** instead.

Theorem 10.14 (Limit Comparison Test).

Similar to comparison test for $\displaystyle\sum_{n=1}^{\infty} a_n$, we may also to construct $\displaystyle\sum_{n=1}^{\infty} b_n$ to compare with. The limit comparison test indicates:

If $0 < \displaystyle\lim_{n\to\infty} \frac{a_n}{b_n} < \infty$, then $\displaystyle\sum_{n=1}^{\infty} a_n$ and $\displaystyle\sum_{n=1}^{\infty} b_n$ both diverge or converge.

If $\displaystyle\lim_{n\to\infty} \frac{a_n}{b_n} = \infty$, then: $\displaystyle\sum_{n=1}^{\infty} b_n$ diverge implies $\displaystyle\sum_{n=1}^{\infty} a_n$ diverge.

If $\displaystyle\lim_{n\to\infty} \frac{a_n}{b_n} = 0$, then: $\displaystyle\sum_{n=1}^{\infty} b_n$ converge implies $\displaystyle\sum_{n=1}^{\infty} a_n$ converge.

Example (8). Determine whether the series $\displaystyle\sum_{n=1}^{\infty} \frac{n^2 + 1}{n^4}$ converges or diverges.

Solution.

Example (9). Determine whether the series $\displaystyle\sum_{n=1}^{\infty} \frac{3n - 1}{2n^3 - 4n + 5}$ converges or diverges.

Solution.

Example (10). Determine whether the series $\displaystyle\sum_{n=1}^{\infty} \frac{\sqrt{n}}{2n^2 + n + 5}$ converges or diverges.

Solution.

Example (11). Determine whether the series $\displaystyle\sum_{n=1}^{\infty} \sin\left(\frac{1}{n}\right)$ converges or diverges.

(*Hint: Consider the limit:* $\displaystyle\lim_{x \to 0} \frac{\sin x}{x} = 1$ *and use the substitution* $x = \frac{1}{n}$)

Solution.

For some series with even more complicated expression such as the powers and factorials. We may use the **ratio test** to determine the convergence.

Theorem 10.15 (Ratio Test).

Let $\sum_{n=1}^{\infty} a_n$ be a series with positive terms, the ratio test is given by the following rules.

Consider the ratio $\rho = \lim_{n\to\infty} \dfrac{a_{n+1}}{a_n}$

If $\rho < 1$, then the series converges.

If $\rho > 1$, then the series diverges.

If $\rho = 1$, then the test is inconclusive (the series might converge or diverge).

Note that, the ratio test is very useful to check the series with powers or factorials.

Example (12). Determine whether the series $\sum_{n=1}^{\infty} \dfrac{3^n}{n^2}$ converges or diverges.

Solution.

Example (13). Determine whether the series $\sum_{n=1}^{\infty} \dfrac{n^3}{(\ln 3)^n}$ converges or diverges.

Solution.

Example (14). Determine whether the series $\dfrac{(1!)^2}{2!} + \dfrac{(2!)^2}{4!} + \cdots + \dfrac{(n!)^2}{(2n)!} + \cdots$ converges or diverges.

Solution.

Theorem 10.16 (Root Test).

Let $\displaystyle\sum_{n=1}^{\infty} a_n$ be a series with positive terms, the root test is used to check the terms in the form of $(f(n))^n$, which is given by the following rules:

If $\displaystyle\lim_{n\to\infty} \sqrt[n]{a_n} < 1$, then the series converges.

If $\displaystyle\lim_{n\to\infty} \sqrt[n]{a_n} > 1$, then the series diverges.

If $\displaystyle\lim_{n\to\infty} \sqrt[n]{a_n} = 1$, then the test is inconclusive (the series might converge or diverge).

Example (15). Determine whether the series $\displaystyle\sum_{n=1}^{\infty} \dfrac{n^n}{2^{3n-1}}$ converges or diverges.

Solution.

Example (16). Determine whether the series $\dfrac{1}{(\ln 2)^2} + \dfrac{1}{(\ln 3)^3} + \dfrac{1}{(\ln 4)^4} + \cdots$ converges or diverges.

Solution.

The integral test for convergence is a method used to test infinite series of monotonous terms for convergence, which can also the most rigorous way to prove the convergence or divergence of infinite series.

Theorem 10.17 (Integral Test).

Let $f(x)$ be a function which is *continuous, positive, and decreasing* for all x in the range $[1, +\infty]$. Then the series:

$$\sum_{n=1}^{\infty} f(n) = f(1) + f(2) + f(3) + \cdots + f(n) + \cdots$$

converges if $\displaystyle\int_{1}^{\infty} f(x)\, \mathrm{d}x$ converges and diverges if $\displaystyle\int_{1}^{\infty} f(x)\, \mathrm{d}x$ diverges.

Example (17). Determine whether the series $\displaystyle\sum_{n=0}^{\infty} ne^{-n}$ converges or diverges.

Solution.

Example (18). Determine whether the series $\displaystyle\sum_{n=2}^{\infty} \frac{1}{n \ln n}$ converges or diverges.

Solution.

Example (19). Use the integral test to show that the p-series $\displaystyle\sum_{n=1}^{\infty} \frac{1}{n^p}$ converges for $p > 1$ and diverges for $p \leq 1$.

Solution.

10.5   Alternating Series Test for Convergence

In the previous SECTIONS, we have introduced several tests for series with *positive terms*. For some series that is alternating with sign, we can use the alternating series test to determine the convergence.

Theorem 10.18 (Alternating Series Test).

A series in which successive terms have opposite signs is called an **alternating series**. For the alternating series in the form of $\sum_{n=1}^{\infty} (-1)^{n+1} a_n$ or $\sum_{n=1}^{\infty} (-1)^n a_n$:

If the series meets the following condition

(1) $a_{n+1} < a_n$ for all possible n in the series.

(2) $\lim_{n \to \infty} a_n = 0$

Then the alternating series converge.

Example (1). Determine whether the series $\sum_{n=1}^{\infty} (-1)^n \frac{2}{3n}$ converges or diverges.

Solution.

Corollary 10.19 (Absolute and Conditional Convergence).

For a convergent alternating series $\sum_{n=1}^{\infty} a_n$, we may check the absolute value that:

If $\sum_{n=1}^{\infty} |a_n|$ also converges, then $\sum_{n=1}^{\infty} a_n$ is absolutely convergent.

If $\sum_{n=1}^{\infty} |a_n|$ diverges, then $\sum_{n=1}^{\infty} a_n$ is conditionally convergent.

* In some cases, we can directly evaluate the convergence for $\sum_{n=1}^{\infty} |a_n|$, then we can deduce:

If $\sum_{n=1}^{\infty} |a_n|$ is tested to be convergent, then $\sum_{n=1}^{\infty} a_n$ must be absolutely convergent.

Example (2). Determine whether the series $\sum_{n=1}^{\infty} (-1)^n \frac{2}{3n}$ is absolutely convergent, conditionally convergent or divergent.

Solution.

Example (3). Determine whether the series $\displaystyle\sum_{n=1}^{\infty} \frac{(-1)^{n+1}}{n!}$ is absolutely convergent, conditionally convergent or divergent.

Solution.

Example (4). Determine whether the series $\displaystyle\sum_{n=1}^{\infty} \frac{(-1)^{n+1}}{5n-1}$ is absolutely convergent, conditionally convergent or divergent.

Solution.

10.6   Power Series and Interval of Convergence

In this sections, we will begin with the concept of **power series**, which is a very power tool to help us make approximations for some functions.

Lemma 10.20 (Power Series).

A series, terms of which are *power functions* of variable x, is called the power series:

$$\sum_{n=0}^{\infty} a_n x^n = a_0 + a_1 x + a_2 x^2 + \cdots + a_n x^n + \cdots$$

Sometimes we may use the term $(x - c)$ as the bases of power functions, which is:

$$\sum_{n=0}^{\infty} a_n (x - c)^n = a_0 + a_1 (x - c) + a_2 (x - c)^2 + \cdots + a_n (x - c)^n + \cdots$$

The power series might converge or diverge depending on the value of x.

Example (1). For a power series $\sum_{n=0}^{\infty} x^n = 1 + x + x^2 + x^3 + \cdots$, find the range of value of x such that the series converge.

Solution.

Theorem 10.21 (Interval and Radius of Convergence).

Consider the function $f(x) = \sum_{n=1}^{\infty} a_n (x - c)^n$. The domain of this function (where it is well-defined) is the set of those values of x such that the series converges.

The domain of such function is called the **interval of convergence**, which is usually written in the form of $(c - R, c + R)$ for some $R > 0$, (together with one or both of the endpoints), the R is called the **radius of convergence**.

Convergence of the series at the endpoints is determined separately.

The fastest way to determine the interval of convergence is the ratio test or geometric series test.

Example (2). Show that $\sum_{n=0}^{\infty} \frac{(x + 3)^n}{n!}$ is convergent for all value of x.

Solution.

Example (3). Determine the radius of convergence for the power series $f(x) = \sum_{n=1}^{\infty} a_n (x - c)^n$.

Solution.

Corollary 10.22 (Determine The Radius of Power Series).

To determine the interval of Convergence, we may first use the **Radius Corollary** that:

The Radius of Convergence of $\sum_{n=1}^{\infty} a_n (x - c)^n$ is $R = \lim_{n \to \infty} \left| \dfrac{a_n}{a_{n+1}} \right|$.

The inequality for checking the convergence (with ratio test) is:

$$\rho = |x - c| \cdot \lim_{n \to \infty} \left| \frac{a_{n+1}}{a_n} \right| < 1 \implies |x - c| < \lim_{n \to \infty} \left| \frac{a_n}{a_{n+1}} \right|$$

Thus, the **center of the interval** is located at $x = c$, and the complete interval of convergence might be:

(1) $c - R < x < c + R$ or   (2) $c - R \le x < c + R$

(3) $c - R < x \le c + R$ or   (4) $c - R \le x \le c + R$

Example (4). Determine the interval of convergence for the power series $\sum_{n=0}^{\infty} nx^n$.

Solution.

Example (5). For what values of x does the series $\displaystyle\sum_{n=0}^{\infty} \frac{x^n}{n+1}$ converge?

Solution.

Example (6). Determine the radius and interval of convergence for the power series $\displaystyle\sum_{n=0}^{\infty} \frac{(-1)^n (x-2)^n}{n^2}$.

Solution.

Corollary 10.23 (Derivatives and Integrals of Power Series).

If f is represented by a ***power series*** centered at c, then:

1. $\displaystyle f'(x) = \left(\sum_{n=0}^{\infty} a_n (x-c)^n \right)' = \sum_{n=0}^{\infty} n \cdot a_n (x-c)^{n-1}$

2. $\displaystyle \int f(x)\, \mathrm{d}x = \int \sum_{n=0}^{\infty} a_n (x-c)^n\, \mathrm{d}x = \sum_{n=0}^{\infty} \frac{a_n (x-c)^{n+1}}{n+1}$

The interval of convergence **is the same as** the original series.

10.7   Maclaurin and Taylor Series

The **Maclaurin and Taylor Series** are series that approximates any function (both polynomial or non-polynomial). They are represented in the power series forms as described in the previous SECTION.

Theorem 10.24 (Maclaurin Series).

The Maclaurin Series is a **power series expansion** of a function, whose center is $x = 0$.

The series, denoted as $P(x)$ must hold the following equality:

For any positive integer n, $P^{(n)}(x) = f^{(n)}(x)$, the general form of the Maclaurin Series for any n-th differentiable function is:

$$f(x) = f(0) + f'(0) \cdot x + \frac{f''(0)}{2!} \cdot x^2 + \frac{f'''(0)}{3!} \cdot x^3 + ... = \sum_{n=0}^{\infty} \frac{f^{(n)}(0)}{n!} \cdot x^n$$

Example (1). For the Maclaurin Series of a function:

$$(x) = a_0 + a_1 x + a_2 x^2 + a_3 x^3 + ...a_n x^n + ... = P(x)$$

Show that for any positive integer n, $a_n = \dfrac{f^{(n)}(0)}{n!}$

Solution.

Example (2). Find the Maclaurin Series for $f(x) = e^x$, and show that the series converges for any real value of x.

Solution.

Besides e^x discussed in the previous EXAMPLE, we can use the similar way to determine the **Maclaurin Series** and the interval of convergence for other functions.

Corollary 10.25 (Some Useful Series Expansions).

$$e^x = 1 + x + \frac{x^2}{2!} + \frac{x^3}{3!} + \cdots = \sum_{n=0}^{\infty} \frac{x^n}{n!} \qquad\qquad -\infty < x < \infty$$

$$\ln(1+x) = x - \frac{x^2}{2} + \frac{x^3}{3} - \frac{x^4}{4} + \cdots = \sum_{n=0}^{\infty} (-1)^{n-1} \frac{x^n}{n} \qquad\qquad -1 < x \le 1$$

$$\frac{1}{1-x} = 1 + x + x^2 + x^3 + x^4 + \cdots = \sum_{n=0}^{\infty} x^n \qquad\qquad -1 < x < 1$$

$$\sin x = x - \frac{x^3}{3!} + \frac{x^5}{5!} - \frac{x^7}{7!} + \cdots = \sum_{n=0}^{\infty} (-1)^n \frac{x^{2n+1}}{(2n+1)!} \qquad\qquad -\infty < x < \infty$$

$$\cos x = 1 - \frac{x^2}{2!} + \frac{x^4}{4!} - \frac{x^6}{6!} + \cdots = \sum_{n=0}^{\infty} (-1)^n \frac{x^{2n}}{(2n)!} \qquad\qquad -\infty < x < \infty$$

Example (3). Find the Maclaurin Series for $f(x) = e^{x^2}$.

Solution.

Example (4). Find the first 3 non-zero terms of the Maclaurin Series for $f(x) = \sin\left(5x + \frac{\pi}{4}\right)$.

Solution.

Example (5). Find the Maclaurin series for $\cos^2 x$ and write it in sigma notation.

Solution.

Theorem 10.26 (Taylor Series).

The Taylor Series is a power series expansion of a function, whose center is $x = c$.

In the same way, we still want $P^{(n)}(x) = f^{(n)}(x)$;

By differentiating both sides, we can see

$$f(c) = P(c) = a_0, \qquad\qquad f'(c) = P'(c) = a_1,$$
$$f''(c) = P''(c) = 2! \cdot a_2, \qquad\qquad f'''(c) = P'''(c) = 3! \cdot a_3 \cdots$$

Therefore, $f^{(n)}(c) = P^{(n)}(c) = n! \cdot a_n \implies a_n = \dfrac{f^{(n)}(c)}{n!}$

The Taylor Series with center $x = c$ is represented by:

$$f(x) = f(c) + f'(c)(x - c) + \frac{f''(c)}{2!}(x - c)^2 + \frac{f'''(c)}{3!} \cdot (x - c)^3 + \dots$$

$$= \sum_{n=0}^{\infty} \frac{f^{(n)}(c)}{n!}(x - c)^n$$

In fact, the **Maclaurin Series** is a special a special **Taylor Series** with $c = 0$.

Example (6). Determine the Taylor series for $f(x) = 3x^2 - 6x + 5$ about the point $x = 1$.

Solution.

162

Lemma 10.27 (Taylor Polynomial).

The k-th degree **Taylor Polynomial** is the k-th order partial sum of the power series:

$$P_k(x) = \sum_{n=0}^{k} \frac{f^{(n)}(c)}{n!} (x - c)^n$$

$$= f(c) + f'(c)(x - c) + \frac{f''(c)}{2!}(x - c)^2 + \dots \frac{f^{(k)}(c)}{k!}(x - c)^k$$

When $c = 0$, the Polynomial is also named as the k-th degree **Maclaurin Polynomial**.

Example (7). Use the 4th degree Maclaurin polynomial for $\ln(1 + x)$ to approximate $\ln(1.1)$.

Solution.

Example (8). The function f is defined by $f(x) = \sin(x^2)$. What are the first four nonzero terms of the Taylor series for f', about $x = 0$?

Solution.

Example (9). The Maclaurin series for the function f is given by $\sum_{n=1}^{\infty} \frac{(-2)^n x^{2n}}{n}$. Find $f^{(4)}(0)$.

Solution.

Example (10). Find the Maclaurin series for $(1 + x)^n$.

Solution.

10.8   Error Bound for Approximations

One of the most important application for power series is approximating funcation values. It gives a solution to the problem of computing difficult quantities: find an easily compute quantity which is sufficiently close to the desired one. The **Error Bound** is an essential criterion to evaluate how *close* your approximation is.

> **Lemma 10.28** (Approximation by Taylor Polynomial).
>
> The Taylor Series of $f(x)$ at $x = c$ is stated as:
>
> $$f(x) = f(c) + f'(c)(x - c) + \frac{f''(c)}{2!}(x - c)^2 + \frac{f'''(c)}{3!} \cdot (x - c)^3 + \cdots$$
>
> When we use the n-th degree Taylor polynomial to approximate the value, we get:
>
> $$f(x) = P_n(x) + R_n(x)$$
>
> Where $P_n(x)$ is the Taylor Polynomial, and $R_n(x)$ is the remainder.
>
> $R_n(x)$ is actually the error of approximation, to estimate the error, we need the **error bound**.

> **Theorem 10.29** (Alternating Series Error Bound).
>
> The error is the difference between partial sum and the limiting value, but by adding an additional term the next partial sum will go past the actual value. Thus, for a series that meets the conditions of the **alternating series test** the error is less than the absolute value of the first omitted term:
>
> $$\text{Error} = |S - S_n| \leq |a_{n+1}|$$

> **Example** (1). If the series $\displaystyle\sum_{n=1}^{\infty} \frac{(-1)^n}{5n + 1}$ is approximated by the partial sum with 15 terms, what is the alternating series error bound?
>
> **Solution.**

> **Example** (2). The function f is defined by the power series $\displaystyle\sum_{n=1}^{\infty} \frac{(-1)^n x^n}{(2n + 1)!}$. Show that the first three terms of the power series approximate $f(1)$ with error less than $\dfrac{1}{4000}$.
>
> **Solution.**

Example (3). Based on the alternating series error bound, what is the least value of m such that:

$$\left| \ln(1.1) - \sum_{n=1}^{m} \frac{(-1)^{n+1}}{n} \cdot (0.1)^n \right| < \frac{1}{5000}$$

Solution.

Theorem 10.30 (Taylor Theorem and Lagrange Error Bound).

The Taylor Theorem indicates that for any function with derivatives through order $n+1$ on an interval containing c, then there must exist a number z within the interval such that:

$$f(x) = \sum_{k=1}^{n} \frac{f^{(k)}(c)}{k!}(x-c)^k + \frac{f^{(n+1)}(z)}{(n+1)!}(x-c)^{n+1}$$

Where the latter them is the remainder.

Kowing that the exact value of z is hard to determine, but we can find the **upper bound** for the remainder. The **Lagrange Error Bound** is such a tool that:

$$\left| \frac{f^{(n+1)}(z)}{(n+1)!}(x-c)^{n+1} \right| \leq \max_{x \in I} \left| f^{(n+1)}(x) \right| \cdot \frac{|x-c|^{n+1}}{(n+1)!}$$

Example (4). Based on the Lagrange error bound, what is the least value of m such that:

$$\left| \ln(1.1) - \sum_{n=1}^{m} \frac{(-1)^{n+1}}{n} \cdot (0.1)^n \right| < \frac{1}{5000}$$

Solution.

Example (5). Let f be a function that has derivatives of all orders for all real numbers and let $P_3(x)$ be the third-degree Taylor Polynomial for f about $x = 0$. $\left|f^{(n)}(x)\right| \leq \dfrac{n}{n+1}$ for all $1 \leq n \leq 5$ and for all values of x. What is the smallest value of k for which the Lagrange Error Bound guarantees that $|f(1) - P_3(1)| \leq k$?

$$\left|\ln(1.1) - \sum_{n=1}^{m} \frac{(-1)^{n+1}}{n} \cdot (0.1)^n\right| < \frac{1}{5000}$$

Solution.

Example (6). The function f has derivatives of all orders for all real numbers. Values of f and its first four derivatives at $x = 2$ are given in the table.

x	$f(x)$	$f'(x)$	$f''(x)$	$f'''(x)$	$f^{(4)}(x)$
2	6	-12	18	-24	34

The fourth derivative of f satisfies the inequality $\left|f^{(4)}(x)\right| \leq 48$ for all $x > 1$. Use the Lagrange error bound to show that the approximation for $f(1.5)$ using the third-degree Taylor Polynomial differs from $f(1.5)$ by no more than $\dfrac{1}{8}$.

Solution.

11   Solutions to the Examples

11.1   Unit 1 Solutions

Example (1.1-1). For the following functions, find the limit of $f(x)$ at $x = 2$.

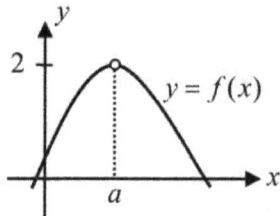 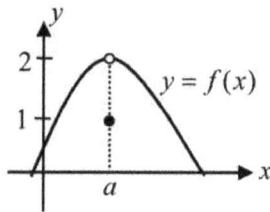 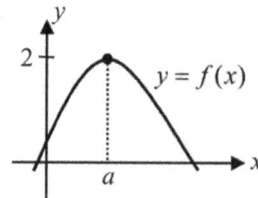

Solution. $\lim\limits_{x \to 2} f(x) = 2$ $\lim\limits_{x \to 2} f(x) = 2$ $\lim\limits_{x \to 2} f(x) = 2$

Example (1.1-2). Determine $\lim\limits_{x \to 0} \dfrac{|x|}{x}$.

Solution.

$\lim\limits_{x \to 0^+} \dfrac{|x|}{x} = \lim\limits_{x \to 0^+} \dfrac{x}{x} = 1$ and $\lim\limits_{x \to 0^-} \dfrac{|x|}{x} = \lim\limits_{x \to 0^-} \dfrac{-x}{x} = -1$. Thus limit does not exist.

Example (1.1-3). If $y = \begin{cases} e^{2x}, & -4 \le x < 0 \\ xe^x, & 0 \le x \le 4 \end{cases}$, find $\lim\limits_{x \to 0} f(x)$.

Solution.

$\lim\limits_{x \to 0^+} f(x) = \lim\limits_{x \to 0^+} xe^x = 0$ and $\lim\limits_{x \to 0^-} f(x) = \lim\limits_{x \to 0^-} e^{2x} = 1$. Thus limit does not exist.

Example (1.2-1). Evaluate $\lim\limits_{x \to 2} \left(5x^2 - 3x + 1 \right)$

Solution. The limit is given by:

$$\lim\limits_{x \to 2} \left(5x^2 - 3x + 1 \right) = 5 \lim\limits_{x \to 2} x^2 - 3 \lim\limits_{x \to 2} x + \lim\limits_{x \to 2} 1$$
$$= 5 \cdot 4 - 3 \cdot 2 + 1 = 15$$

Example (1.2-2). Evaluate $\lim\limits_{x \to \pi} 3x \cdot \sin x$

Solution. The limit is given by:

$$\lim\limits_{x \to \pi} 3x \cdot \sin x = 3 \left(\lim\limits_{x \to \pi} x \right) \cdot \left(\lim\limits_{x \to \pi} \sin x \right) = 3\pi \cdot \sin \pi = 0$$

Example (1.2-3). Evaluate $\lim\limits_{x \to 1} \dfrac{3x^2 + 2x - 1}{x^2 + 1}$

Solution. The limit is given by:

$$\lim_{x \to 1} \frac{3x^2 + 2x - 1}{x^2 + 1} = \frac{\lim_{x \to 1} \left(3x^2 + 2x - 1\right)}{\lim_{x \to 1} \left(x^2 + 1\right)}$$
$$= \frac{3 + 2 - 1}{1^2 + 1} = 2$$

Example (1.2-4). Evaluate $\lim_{t \to 2} \dfrac{t^2 - 3t + 2}{t - 2}$

Solution. The limit is given by:

$$\lim_{t \to 2} \frac{t^2 - 3t + 2}{t - 2} = \lim_{t \to 2} \frac{(t - 1)(t - 2)}{t - 2} = \lim_{t \to 2} (t - 1) = 1$$

Example (1.2-5). Evaluate $\lim_{x \to b} \dfrac{x^5 - b^5}{x^{10} - b^{10}}$

Solution. The limit is given by:

$$\lim_{x \to b} \frac{x^5 - b^5}{x^{10} - b^{10}} = \lim_{x \to b} \frac{x^5 - b^5}{(x^5 + b^5)(x^5 - b^5)} = \lim_{x \to b} \frac{1}{x^5 + b^5} = \frac{1}{2b^5}$$

Example (1.2-6). Evaluate $\lim_{x \to -2} \dfrac{x^3 + 8}{x^2 - 4}$

Solution. The limit is given by:

$$\lim_{x \to -2} \frac{x^3 + 8}{x^2 - 4} = \lim_{x \to -2} \frac{(x + 2)(x^2 - 2x + 4)}{(x + 2)(x - 2)} = \lim_{x \to -2} \frac{x^2 - 2x + 4}{x - 2} = \frac{4 - (-4) + 4}{-4} = -3$$

Example (1.2-7). Evaluate $\lim_{t \to 0} \dfrac{\sqrt{t + 2} - \sqrt{2}}{t}$

Solution. The limit is given by:

$$\lim_{t \to 0} \frac{\sqrt{t + 2} - \sqrt{2}}{t} \cdot \frac{\sqrt{t + 2} + \sqrt{2}}{\sqrt{t + 2} + \sqrt{2}} = \lim_{t \to 0} \frac{t + 2 - 2}{t(\sqrt{t + 2} + \sqrt{2})} = \lim_{t \to 0} \frac{t}{t(\sqrt{t + 2} + \sqrt{2})}$$
$$= \lim_{t \to 0} \frac{1}{\sqrt{t + 2} + \sqrt{2}} = \frac{1}{\sqrt{0 + 2} + \sqrt{2}} = \frac{1}{2\sqrt{2}} = \frac{\sqrt{2}}{4}$$

Example (1.2-8). Evaluate $\lim_{x \to 1} \dfrac{x - 1}{\sqrt{x + 3} - 2}$

Solution. The limit is given by:

$$\lim_{x \to 1} \frac{(x-1)(\sqrt{x+3}+2)}{(\sqrt{x+3}-2)(\sqrt{x+3}+2)} = \lim_{x \to 1} \frac{(x-1)(\sqrt{x+3}+2)}{x-1} = \lim_{x \to 1} \sqrt{x+3}+2 = 4$$

Example (1.2-9). Evaluate $\lim_{x \to 0} \left(x^2 \sin \left(\frac{\pi}{x} \right) \right)$

Solution.

Since $-1 \leq \sin \left(\frac{\pi}{x} \right) \leq 1$, then $-x^2 \leq x^2 \sin \left(\frac{\pi}{x} \right) \leq x^2$.

Knowing that $\lim_{x \to 0} -x^2 = 0$ and $\lim_{x \to 0} x^2 = 0$, by the **Squeeze Theorem**, $\lim_{x \to 0} \left(x^2 \sin \left(\frac{\pi}{x} \right) \right) = 0$

Example (1.2-10). Evaluate $\lim_{x \to 0} \frac{\sin 3x}{x}$

Solution. The limit is given by:

$$\lim_{x \to 0} \frac{\sin 3x}{x} = \lim_{x \to 0} \frac{3 \sin 3x}{3x} = 3 \cdot \lim_{x \to 0} \frac{\sin 3x}{3x} = 3 \cdot \lim_{3x \to 0} \frac{\sin 3x}{3x} = 3$$

Example (1.2-11). Evaluate $\lim_{y \to 0} \frac{y^2}{1 - \cos y}$

Solution. The limit is given by:

$$\lim_{y \to 0} \frac{y^2}{1 - \cos y} = \lim_{y \to 0} \frac{y^2}{1 - \cos y} \cdot \frac{1 + \cos y}{1 - \cos y}$$

$$= \lim_{y \to 0} \frac{y^2 (1 + \cos y)}{1 - \cos^2 y} = \lim_{y \to 0} \frac{y^2 (1 + \cos y)}{\sin^2 y}$$

$$= \lim_{y \to 0} \left(\frac{y}{\sin y} \right)^2 \cdot \lim_{y \to 0} (1 + \cos y) = 1^2 \cdot (1 + 1) = 2$$

Note that:

$$\lim_{y \to 0} \frac{y}{\sin y} = \lim_{y \to 0} \frac{1}{\frac{\sin y}{y}} = \frac{1}{1} = 1$$

Example (1.2-12). Evaluate the following limit: $\lim_{x \to 2} f(g(x))$, by the following graph.

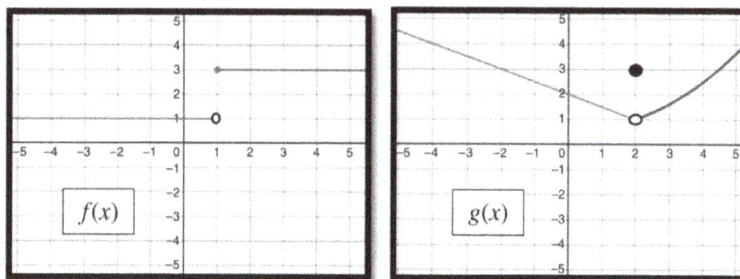

Solution.

As x approach 2 **from the left**, $\lim\limits_{x \to 2^-} g(x) = 1$, but it approaches from **above**, thus:

$$\lim_{x \to 2^-} f(g(x)) = \lim_{x \to 1^+} f(x) = 3$$

As x approach 2 **from the right**, $\lim\limits_{x \to 2^+} g(x) = 1$, but it approaches from **above**, thus:

$$\lim_{x \to 2^+} f(g(x)) = \lim_{x \to 1^+} f(x) = 3$$

Since $\lim\limits_{x \to 2^-} f(g(x)) = \lim\limits_{x \to 2^+} f(g(x)) = 3$, then $\lim\limits_{x \to 2} f(g(x)) = 3$.

Example (1.2-13).

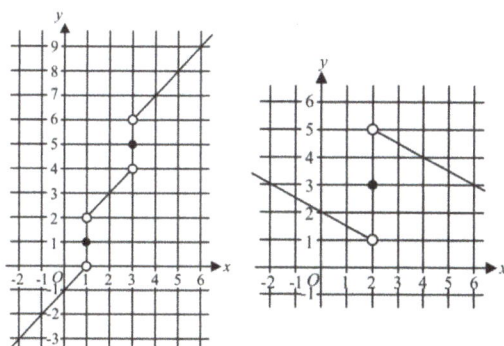

Graph of f   Graph of g

The graphs of the functions f and g are shown. What is $\lim\limits_{x \to 2^-} f(g(x))$?

Solution. $\lim\limits_{x \to 2^-} g(x) = 1$, $\lim\limits_{x \to 2^-} f(g(x)) = \lim\limits_{x \to 1^+} f(x) = 2$

Example (1.3-1). Evaluate the limits: (a) $\lim\limits_{x \to 2^+} \dfrac{3x-1}{x-2}$ and (b) $\lim\limits_{x \to 2^-} \dfrac{3x-1}{x-2}$

Solution.

(a) The limit of the numerator is 5 and the limit of the denominator is 0 through positive values.

$$\text{Thus, } \lim_{x \to 2^+} \frac{3x-1}{x-2} = \infty$$

(b) The limit of the numerator is 5 and the limit of the denominator is 0 through negative values.

$$\text{Thus, } \lim_{x \to 2^-} \frac{3x-1}{x-2} = -\infty$$

Example (1.3-2). Evaluate $\displaystyle\lim_{x\to 5^-} \frac{\sqrt{25-x^2}}{x-5}$

Solution. The limit is given by:

$$\lim_{x\to 5^-}\frac{\sqrt{25-x^2}}{x-5} = -\lim_{x\to 5^-}\frac{\sqrt{(5-x)(5+x)}}{\sqrt{(5-x)(5-x)}} = -\lim_{x\to 5^-}\sqrt{\frac{5+x}{5-x}} = -\infty$$

Example (1.3-3). Evaluate $\displaystyle\lim_{x\to 2^-}\frac{\lfloor x\rfloor - x}{2-x}$, where $\lfloor x\rfloor$ is the greatest integer value of x.

Solution.

As $x\to 2^-$, $\lfloor x\rfloor = 1$. The limit of the numerator is $1-2 = -1$

As $x\to 2^-$, $2-x = 0$, through positive values.

Thus, $\displaystyle\lim_{x\to 2^-}\frac{\lfloor x\rfloor - x}{2-x} = -\infty$

Example (1.3-4). Evaluate $\displaystyle\lim_{x\to\infty}\frac{6x-13}{2x+5}$.

Solution. Since $\delta(p) = \delta(q)$, the limit is given by:

$$\lim_{x\to\infty}\frac{6x-13}{2x+5} = \frac{6}{2} = 3$$

Example (1.3-5). Evaluate $\displaystyle\lim_{x\to\infty}\frac{2x+1}{\sqrt{x^2+3}}$.

Solution. Since $\delta(p) = \delta(q)$, the limit is given by:

$$\lim_{x\to\infty}\frac{2x+1}{\sqrt{x^2+3}} = \lim_{x\to\infty}\frac{2x+1}{|x|\cdot\sqrt{1+\frac{3}{x^2}}} = \lim_{x\to\infty}\frac{2x+1}{-x\cdot\sqrt{1+\frac{3}{x^2}}} = -2$$

Example (1.3-6). Evaluate $\displaystyle\lim_{x\to\infty} \frac{1-x^2}{10x+7}$.

Solution. Since $\delta(p) > \delta(q)$, the limit is given by:

$$\lim_{x\to\infty} \frac{1-x^2}{10x+7} = -\infty$$

Note that $1-x^2$ is negative and $10x+7$ is positive when x is very large, so the ratio must be negative.

Example (1.3-7). Find the horizontal and vertical asymptotes of the function $f(x) = \dfrac{3x+5}{x-2}$.

Solution.

Horizontal: $\displaystyle\lim_{x\to-\infty} \frac{3x+5}{x-2} = \lim_{x\to\infty} \frac{3x+5}{x-2} = 3$. Thus, $y=3$ is a horizontal asymptote.

Vertical: check the denominator $x - 2 = 0 \Rightarrow x = 2$

$$\lim_{x\to 2^-} \frac{3x+5}{x-2} = -\infty, \quad \lim_{x\to 2^+} \frac{3x+5}{x-2} = \infty, \quad \text{Thus, } x = 2 \text{ is a vertical asymptote.}$$

Example (1.3-8). Find the horizontal and vertical asymptotes of the function $f(x) = \dfrac{2e^x - 1}{3 - 5e^x}$.

Solution.

Horizontal: We check the limits when $x \to \pm\infty$

$$\lim_{x\to\infty} \frac{2e^x-1}{3-5e^x} = \lim_{x\to\infty} \frac{2-\frac{1}{e^x}}{\frac{3}{e^x}-5} = -\frac{2}{5} \quad \text{and} \quad \lim_{x\to-\infty} \frac{2e^x-1}{3-5e^x} = -\frac{1}{3},$$

Thus, $y = -\dfrac{2}{5}$ and $y = -\dfrac{1}{3}$ are horizontal asymptotes

Vertical: check the denominator $3 - 5e^x = 0 \Rightarrow e^x = \dfrac{3}{5} \Rightarrow x = \ln\dfrac{3}{5}$

$$\lim_{x\to\ln\frac{3}{5}^-} \frac{2e^x-1}{3-5e^x} = \infty, \quad \lim_{x\to\ln\frac{3}{5}^+} \frac{2e^x-1}{3-5e^x} = -\infty, \quad \text{Thus, } x = \ln\frac{3}{5} \text{ is a vertical asymptote.}$$

Example (1.3-9). Using your calculator, find the horizontal asymptotes of the function $f(x) = \dfrac{2\sin x}{x}$.

Solution.

The graph shows that $f(x)$ oscillates back and forth about the x-axis.
As $x \to \pm\infty$, which implies that $f(x)$ approaches 0.
Thus, the line $y = 0$ (or the x-axis) is a horizontal asymptote.

Example (1.4-1). Find the points of discontinuity of the function $f(x) = \dfrac{x+5}{x^2 - x - 2}$

Solution.

The function is not defined when the denominator is 0.

$$x^2 - x - 2 = 0$$
$$(x - 2)(x + 1) = 0 \Rightarrow x = 2 \text{ or } x = -1$$

The function $f(x)$ is discontinuous at $x = -1$ and at $x = 2$.

Example (1.4-2). Determine the intervals on which the given function is continuous:

$$f(x) = \begin{cases} \dfrac{x^2 + 3x - 10}{x - 2}, & x \neq 2 \\ 10, & x = 2 \end{cases}$$

Solution.

Let's check the three conditions of continuity at $x = 2$:

1. $\displaystyle\lim_{x \to 2} \dfrac{x^2 + 3x - 10}{x - 2} = \lim_{x \to 2} \dfrac{(x + 5)(x - 2)}{x - 2} = \lim_{x \to 2} (x + 5) = 7.$

2. $f(2)$ is defined, $f(2) = 10.$

3. $f(2) \neq \displaystyle\lim_{x \to 2} f(x)$

Thus, $f(x)$ is discontinuous at $x = 2$.

Example (1.4-3). For what value of k is the function $f(x) = \begin{cases} \dfrac{5\sin x}{x}, & x < 0 \\ k - 3x, & x \geq 0 \end{cases}$ continuous at $x = 2$?

Solution. $\displaystyle\lim_{x \to 0^-} \dfrac{5\sin x}{x} = 5$ and $f(0) = k$. Thus, $k = 5$.

Example (1.4-4). For what value of k is the function $g(x) = \begin{cases} \dfrac{x^2 - k^2}{x - k}, & x \neq k \\ 10, & x = k \end{cases}$ continuous at $x = k$?

Solution.

The limit of $g(x)$ at $x = k$ is:

$$\lim_{x \to k} \dfrac{x^2 - k^2}{x - k} = \lim_{x \to k} \dfrac{(x + k)(x - k)}{x - k}$$
$$= \lim_{x \to k} (x + k) = 2k$$

Since the limit exists and $f(k)$ is defined by $f(k) = 10$. The function is continuous if and only if:

$$\lim_{x \to k} \dfrac{x^2 - k^2}{x - k} = f(k) \Rightarrow 2k = 10 \Rightarrow k = 5.$$

Example (1.4-5). A function f is continuous on $[0,5]$, and some of the values of f are shown below.

x	0	3	5
f	-4	b	-4

If $f(x) = -2$ has no solution on $[0,5]$, then b could be:

(A) 1   (B) 0   (C) -2 (D) -5

Solution.

If $b = -2$, then $x = 3$ would be a solution for $f(x) = -2$

If $b = 0$, -1, or 3, $f(x) = -2$ would have two solutions for $f(x) = -2$. Thus, $b = -5$.

The answer is (D).

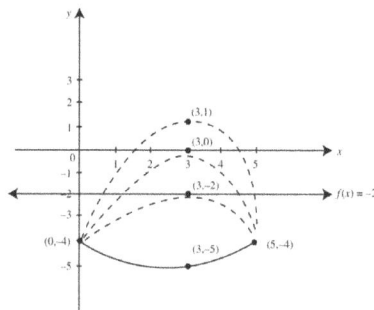

Example (1.4-6). The function f is continuous on the closed interval $[-5,5]$ and have values that are given in the table below. What is the minimum number of times that the function takes on the value $f(x) = -2$ on the interval $[-5,5]$?

x	-5	-3	-1	1	3	5
$f(x)$	4	2	-3	1	-4	-5

(A) 1   (B) 2   (C) 3   (D) 4

Solution.

In the interval $[-3,-1]$, $[-1,1]$ and $[1,3]$, the function $f(x)$ attains values greater than and less than -2. Therefore, by the Intermediate Value Theorem $f(x)$ reaches -2 at least once in each of those intervals.

The answer is (C).

11.2 Unit 2 Solutions

Example (2.1-1). If $f(x) = x^2 - 2x - 3$, find: (a) $f'(x)$ using the definition of derivative,   (b) $f'(0)$, (c) $f'(1)$, and (d) $f'(3)$.

Solution.

(a) Using the definition of derivative, $f'(x) = \lim\limits_{h \to 0} \dfrac{f(x+h) - f(x)}{h}$

$$= \lim_{h \to 0} \frac{[(x+h)^2 - 2(x+h) - 3] - [x^2 - 2x - 3]}{h}$$

$$= \lim_{h \to 0} \frac{(x^2 + 2xh + h^2 - 2x - 2h - 3) - (x^2 - 2x - 3)}{h}$$

$$= \lim_{h \to 0} \frac{2xh + h^2 - 2h}{h}$$

$$= \lim_{h \to 0} (2x + h - 2)$$

$$= 2x - 2$$

(b) $f'(0) = 2(0) - 2 = -2$

(c) $f'(1) = 2(1) - 2 = 0$

(d) $f'(3) = 2(3) - 2 = 4$

Example (2.1-2). Using the definition of derivative to find the derivative of $f(x) = \dfrac{1}{x - 1}$.

Solution.

$$f'(x) = \lim_{h \to 0} \frac{\frac{1}{x+h-1} - \frac{1}{x-1}}{h} = \lim_{h \to 0} \frac{\frac{(x-1)-(x+h-1)}{(x+h-1)(x-1)}}{h}$$

$$= \lim_{h \to 0} \frac{-h}{h(x+h-1)(x-1)}$$

$$= \lim_{h \to 0} \frac{-1}{(x+h-1)(x-1)}$$

$$= -\frac{1}{(x-1)^2}$$

Example (2.1-3). Using the definition of derivative to find the derivative of a quadratic function $f(x) = ax^2 + bx + c$.

Solution.

$$f'(x) = \lim_{h \to 0} \frac{[a(x+h)^2 + b(x+h) + c] - [ax^2 + bx + c]}{h}$$

$$= \lim_{h \to 0} \frac{(ax^2 + 2axh + ah^2 + bx + bh + c) - (ax^2 + bx + c)}{h}$$

$$= \lim_{h \to 0} \frac{2axh + ah^2 + bh}{h}$$

$$= \lim_{h \to 0} (2ax + ah + b)$$

$$= 2ax + b$$

** We will discuss the *formulae and rules of differentiation* later.

Example (2.1-4). Evaluate the limit $\displaystyle\lim_{h \to 0} \frac{\cos(\pi + h) - \cos\pi}{h}$.

Solution.

We may consider two methods:

Solution 1: The expression $\displaystyle\lim_{h \to 0} \frac{\cos(\pi + h) - \cos\pi}{h}$ is equivalent to the derivative of the function $f(x) = \cos x$ at $x = \pi$, ie. $f'(\pi)$. Te derivative of $f(x) = \cos x$ at $x = \pi$ is equivalent to the slope of the tangent to the curve of $\cos x$ at $x = \pi$. Since the tangent is parallel to the x-axis. Thus the slope is 0.

$$\lim_{h \to 0} \frac{\cos\pi + h - \cos\pi}{h} = 0$$

Solution 2: Consider the compound angle formula: $\cos(a + b) = \cos a \cos b - \sin a \sin b$, then:

$$\lim_{h \to 0} \frac{\cos(\pi + h) - \cos\pi}{h} = \lim_{h \to 0} \frac{\cos\pi \cos h - \sin pi \sin h - \cos\pi}{h}$$

$$= \lim_{h \to 0} \frac{-\cos h + 1}{h} = \lim_{h \to 0} \frac{1 - \cos h}{h}$$

$$= \lim_{h \to 0} \frac{(1 - \cos h)(1 + \cos h)}{h(1 + \cos h)}$$

$$= \lim_{h \to 0} \frac{1 - \cos^2 h}{h(1 + \cos h)} = \lim_{h \to 0} \frac{\sin^2 h}{h(1 + \cos h)}$$

$$= \lim_{h \to 0} \frac{\sin h}{h} \cdot \lim_{h \to 0} \frac{\sin h}{(1 + \cos h)} = 0$$

Example (2.2-1). The Graph of a function is defined on the interval $[-6, 12]$. Find every -value at which the function is not differentiable on $(-6, 12)$.

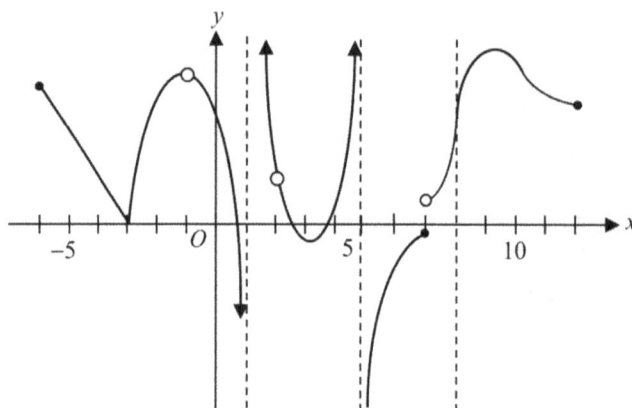

Solution. There are 7 values of x at which the function is not differentiable.

1. **Discontinuity**: $x = -1$, $x = 1$, $x = 2$, $x = 5$, $x = 7$

2. **Vertical Tangent**: $x = 8$

3. **Sharp Corner**: $x = -3$

Example (2.2-2). Determine continuity and differentiability of $f(x) = |2x|$.

Solution.

Continuity:

$$f(x) = \begin{cases} 2x, & x \geq 0 \quad \to f(0) = 0 \\ -2x, & x < 0 \quad \to \lim_{x \to 0^-} (-2x) = 0 \end{cases} \Rightarrow f \text{ is continuous at } x = 0.$$

Differentiability:

$$f'(x) = \begin{cases} 2, & x \geq 0 \quad \to \lim_{x \to 0^+} f'(x) = 1 \\ -2, & x < 0 \quad \to \lim_{x \to 0^-} f'(x) = -1 \end{cases} \Rightarrow \text{not equal, } f \text{ is not differentiable at } x = 0.$$

Example (2.2-3). Determine continuity and differentiability of $f(x) = x^2 - 3|x| + 2$.

Solution.

Continuity:

$$f(x) = \begin{cases} x^2 - 3x + 2, & x \geq 0 \quad \to f(0) = 2 \\ x^2 + 3x + 2, & x < 0 \quad \to \lim_{x \to 0^-} (x^2 + 3x + 2) = 2 \end{cases} \Rightarrow f \text{ is continuous at } x = 0.$$

Differentiability:

$$f'(x) = \begin{cases} 2x - 3, & x \geq 0 \quad \to \lim_{x \to 0^+} 2x - 3 = -3 \\ 2x + 3, & x < 0 \quad \to \lim_{x \to 0^-} 2x + 3 = 3 \end{cases}$$

Since the one sided derivatives are not equal, f is not differentiable at $x = 0$.

Example (2.2-4). Determine whether the function is differentiable at $x = 1$.

(a) $f(x) = \sqrt{1 - x^2}$

(b) $f(x) = \begin{cases} x + 1, & x \leq 1 \\ x^2 + 1, & x > 1 \end{cases}$

Solution.

(a) Since f is defined on the interval $[-1, 1]$, f is not differentiable at $x = 1$.

(b) For $f(x) = \begin{cases} x + 1, & x \leq 1 \\ x^2 + 1, & x > 1 \end{cases}$

$\lim\limits_{x \to 1^+} f(x) = 2$, and $f(1) = 2$, $f(x)$ is continuous at $x = 1$

$$f'(x) = \begin{cases} \lim\limits_{x \to 1^+} \dfrac{f(x) - f(1)}{x - 1} = \lim\limits_{x \to 1^+} \dfrac{x^2 + 1 - 2}{x - 1} = \lim\limits_{x \to 1^+} \dfrac{x^2 - 1}{x - 1} = \lim\limits_{x \to 1^+} (x + 1) = 2, & x \leq 1 \\ \lim\limits_{x \to 1^-} \dfrac{f(x) - f(1)}{x - 1} = \lim\limits_{x \to 1^+} \dfrac{x + 1 - 2}{x - 1} = \lim\limits_{x \to 1^+} \dfrac{x - 1}{x - 1} = \lim\limits_{x \to 1^+} 1 = 1, & x > 1 \end{cases}$$

Since one sided derivatives are not equal, f is not differentiable at $x = 1$.

Example (2.3-1). If $f(x) = 2x^3$, find $f'(x)$.

Solution. $f'(x) = 3 \cdot 2x^2 = 6x^2$

Example (2.3-2). If $y = \dfrac{1}{x^2}$, find $\dfrac{dy}{dx}$ and $\dfrac{dy}{dx}\Big|_{x=0}$.

Solution.

$y = \dfrac{1}{x^2} = x^{-2} \quad \Rightarrow \quad \dfrac{dy}{dx} = -2x^{-3} = -\dfrac{2}{x^3} \qquad$ when $x = 0$, $\dfrac{dy}{dx}\Big|_{x=0} = -\dfrac{2}{0} \to$ undefined

Therefore, $\dfrac{dy}{dx}\Big|_{x=0}$ does not exist.

Example (2.3-3). Find $f'(x)$ and $f'(3)$ if $f(x) = \dfrac{1}{\sqrt{x}}$.

Solution.

$f(x) = \dfrac{1}{\sqrt{x}} = x^{-\frac{1}{2}}$

By the Power Rule: $f'(x) = -\dfrac{1}{2}x^{-\frac{3}{2}} = -\dfrac{1}{2\sqrt{x^3}}$, then $f'(3) = -\dfrac{1}{2\sqrt{27}} = -\dfrac{1}{6\sqrt{3}} = -\dfrac{\sqrt{3}}{18}$

Example (2.3-4). Find $f'(x)$ if $f(x) = x^3 - 10x + 5$.

Solution. $f'(x) = 3x^2 - 10$

Example (2.3-5). If $y = (3x - 5)(x^4 + 8x - 1)$, find $\dfrac{dy}{dx}$.

Solution. By the Product Rule:

$$\begin{aligned}
\frac{dy}{dx} &= (3)\left(x^4 + 8x - 1\right) + \left(4x^3 + 8\right)(3x - 5) \\
&= \left(3x^4 + 24x - 3\right) + \left(12x^4 - 20x^3 + 24x - 40\right) \\
&= 15x^4 - 20x^3 + 48x - 43
\end{aligned}$$

Example (2.3-6). If $f(x) = \dfrac{2x - 1}{x + 5}$, find $f'(x)$.

Solution. By the Quotient Rule:

$$\begin{aligned}
f'(x) &= \frac{(2)(x + 5) - (1)(2x - 1)}{(x + 5)^2} \\
&= \frac{2x + 10 - 2x + 1}{(x + 5)^2} \\
&= \frac{11}{(x + 5)^2}, \quad x \neq -5
\end{aligned}$$

Example (2.4-1). Use the *change-of-base formula* to show that the derivative of $y = \log_2 x$ is $\dfrac{1}{x \ln 2}$.

Solution. By the change-of-base formula:

$$\begin{aligned}
\frac{d}{dx}\left[\log_2 x\right] &= \frac{d}{dx}\left(\frac{\ln x}{\ln 2}\right) \\
&= \frac{1}{\ln 2} \cdot \frac{d}{dx}\left[\ln x\right] = \frac{1}{x \ln 2}
\end{aligned}$$

Example (2.4-2). Use the quotient rule to show that the derivative of $y = \tan x$ is $\sec x \tan x$.

Solution. By the Quotient Rule:

$$\begin{aligned}
\frac{d}{dx}\left[\tan x\right] = \frac{d}{dx}\left(\frac{\sin x}{\cos x}\right) &= \frac{\cos x \cdot \cos x - (-\sin x) \cdot \sin x}{\cos^2 x} \\
&= \frac{\cos^2 x + \sin^2 x}{\cos^2 x} \\
&= \frac{1}{\cos^2 x} = \sec^2 x
\end{aligned}$$

Example (2.4-3). If $f(x) = e^x \cos x$, find $f'(0)$.

Solution. The derivative of $f(x)$ is given by:

$$f'(x) = e^x \cos x + e^x \cdot (-\sin x)$$
$$= e^x(\cos x - \sin x)$$
$$f'(0) = e^0(\cos 0 - \sin 0) = 1 \cdot (1 - 0) = 1$$

Example (2.4-4). If $f(x) = \dfrac{\sin x}{x^2}$, find $f'(x)$.

Solution. The derivative of $f(x)$ is given by:

$$f'(x) = \frac{x^2 \cos x - \sin x \cdot 2x}{x^4} = \frac{x \cos x - 2 \sin x}{x^3}$$

Example (2.4-5). If $y = \dfrac{\tan x}{1 + \tan x}$, find $\dfrac{dy}{dx}$.

Solution. The derivative is given by:

$$\frac{dy}{dx} = \frac{\sec^2 x(1 + \tan x) - \sec^2 x \cdot \tan x}{(1 + \tan x)^2} = \frac{\sec^2 x}{(1 + \tan x)^2}$$

$$= \frac{\dfrac{1}{\cos^2 x}}{(1 + \dfrac{\sin x}{\cos x})^2} = \frac{\dfrac{1}{\cos^2 x}}{\left(\dfrac{\cos x + \sin x}{\cos x}\right)^2} = \frac{1}{(\cos x + \sin x)^2}$$

Example (2.4-6). If $y = 5xe^3 + x^2 e^x$, find $\dfrac{dy}{dx}$.

Solution. Note that $5e^3$ is a constant. The derivative of $f(x)$ is:

$$\frac{dy}{dx} = 5e^3 + 2xe^x + x^2 e^x = \left(x^2 + 2x\right)e^x + 5e^3$$

Example (2.4-7). If $y = \dfrac{\ln x}{e^x}$, find $\dfrac{dy}{dx}$.

Solution. The derivative of $f(x)$ is:

$$\frac{dy}{dx} = \frac{e^x \cdot \frac{1}{x} - \ln x \cdot e^x}{e^{2x}} = \frac{e^x - x \ln x \cdot e^x}{xe^{2x}} = \frac{1 - x \ln x}{xe^x}$$

Example (2.5-1). Find the equation of the tangent line to the curve $y = x \ln x$ at $x = e$.

Solution.

The derivative of the function is: $\dfrac{dy}{dx} = 1 \cdot \ln x + x \cdot \dfrac{1}{x} = \ln x + 1$

when $x = e$, $y = e$, $m = \dfrac{dy}{dx}\bigg|_{x=e} = \ln e + 1 = 2$.

The tangent line passes through the point (e, e) is:

$$y - e = 2(x - e) \ \Rightarrow \ y = 2x - e$$

Example (2.5-2). Write an equation for each normal to the graph of $y = 2 \sin x$ for $0 \le x \le 2\pi$ that has a slope of $\dfrac{1}{2}$.

Solution.

The derivative of the function is: $\dfrac{dy}{dx} = 2 \cos x$

If $m_{\text{normal}} = \dfrac{1}{2}$, then $m_{\text{tangent}} = 2 \cos x = -2$

which implies $\cos x = -1$, thus $x = \pi$, then $y = 2 \sin \pi = 0$, the point is $(\pi, 0)$

Since $m_{\text{normal}} = \dfrac{1}{2}$, the equation of the normal is: $y = \dfrac{1}{2}(x - \pi) = \dfrac{1}{2}x - \dfrac{\pi}{2}$

Example (2.5-3). Find all points on the graph of $y = |xe^x|$ at which the graph has a horizontal tangent.

Solution.

Since $y = \begin{cases} xe^x, & x \ge 0 \\ -xe^x, & x < 0 \end{cases}$ $\Rightarrow$ $\dfrac{dy}{dx} = \begin{cases} e^x + xe^x, & x \ge 0 \\ -e^x - xe^x, & x < 0 \end{cases}$

If the graph has a horizontal tangent, then $\dfrac{dy}{dx} = 0$.

If $x \ge 0$, set $e^x + xe^x = 0 \ \Rightarrow \ e^x(1 + x) = 0 \ \Rightarrow \ x = -1$, which contradicts with $x \ge 0$.

If $x < 0$, set $-e^x - xe^x = 0 \ \Rightarrow \ -e^x(1 + x) = 0 \ \Rightarrow \ x = -1$

At $x = -1$, $y = -(-1) \cdot e^{-1} = \dfrac{1}{e}$.

Thus at the point $\left(1, \dfrac{1}{e}\right)$, the graph has a horizontal tangent.

11.3   Unit 3 Solutions

Example (3.1-1). If $f(x) = (3x - 5)^{10}$, find $\dfrac{dy}{dx}$.

Solution.

Using the **Chain Rule**, let $u = 3x - 5$, then $y = u^{10}$.

Then, $\dfrac{dy}{du} = 10u^9$, and $\dfrac{du}{dx} = 3$. Thus, $\dfrac{dy}{dx} = \dfrac{dy}{du} \times \dfrac{du}{dx} = 10u^9 \cdot 3 = 30\,(3x - 5)^9$.

Example (3.1-2). Find $f'(x)$ if $f(x) = \cot(4x - 6)$.

Solution.

Using the **Chain Rule**, let $u = 4x - 6$, then $y = \cot u$.

Then, $\dfrac{dy}{dx} = 4 \cdot (-\csc^2 u) = -4\csc^2(4x - 6)$.

Example (3.1-3). Find $f'(x)$ if $f(x) = e^{-2x}$.

Solution.

Using the **Chain Rule**, let $u = -2x$, then $y = e^u$.

Then, $\dfrac{dy}{dx} = -2e^u = -2e^{-2x}$.

Example (3.1-4). Find $f'(x)$ if $f(x) = e^{x^3 + 3}$.

Solution.

Using the **Chain Rule**, let $u = x^3 + 3$, then $y = e^u$.

In this problem, we cannot use the previous COROLLARY since u cannot be written in linear form.

Since $\dfrac{dy}{du} = e^u$, and $\dfrac{du}{dx} = 3x^2$

Then, $\dfrac{dy}{dx} = \dfrac{dy}{du} \times \dfrac{du}{dx} = e^u \cdot 3x^2 = 3x^2 e^{x^3 + 3}$.

Example (3.1-5). If $y = 3^{\sin x}$, find $\dfrac{dy}{dx}$.

Solution.

Let $u = \sin x$, then $y = 3^u$.

Since, $\dfrac{dy}{du} = 3^u \cdot \ln 3$, and $\dfrac{du}{dx} = \cos x$.

Thus, $\dfrac{dy}{dx} = \dfrac{dy}{du} \times \dfrac{du}{dx} = 3^u \cdot \ln 3 \cdot \cos x = \ln 3 \cdot \cos x \cdot 3^{\sin x}$.

Example (3.1-6). If $y = (\ln x)^5$, find $\dfrac{dy}{dx}$.

Solution.

Let $u = \ln x$, then $y = u^5$.

Since, $\dfrac{dy}{du} = 5u^4$, and $\dfrac{du}{dx} = \dfrac{1}{x}$. Thus, $\dfrac{dy}{dx} = \dfrac{dy}{du} \times \dfrac{du}{dx} = 5u^4 \cdot \dfrac{1}{x} = \dfrac{5\,(\ln x)^4}{x}$.

Example (3.1-7). If $y = \log_5 \left(x^2 + 2x - 3\right)$, find $\dfrac{dy}{dx}$.

Solution.

Let $u = x^2 + 2x - 3$, then $y = \log_5 u$.

Since, $\dfrac{dy}{du} = \dfrac{1}{u \ln 5}$, and $\dfrac{du}{dx} = 2x + 2$.

Thus, $\dfrac{dy}{dx} = \dfrac{dy}{du} \times \dfrac{du}{dx} = \dfrac{1}{u \ln 5} \cdot (2x + 2) = \dfrac{2x + 2}{(x^2 + 2x - 3) \ln 5}$.

Example (3.1-8). If $f(x) = 5x\sqrt{25 - x^2}$, find $f'(x)$.

Solution.

Rewrite $f(x) = 5x\sqrt{25 - x^2}$ as $f(x) = 5x\left(25 - x^2\right)^{\frac{1}{2}}$

Before we use the product rule, we need to find the derivative of $\left(25 - x^2\right)^{\frac{1}{2}}$.

Let $u = 25 - x^2$, then $\left(25 - x^2\right)^{\frac{1}{2}} = u^{\frac{1}{2}}$

Since $\dfrac{d}{du}\, u^{\frac{1}{2}} = \dfrac{1}{2}u^{-\frac{1}{2}} = \dfrac{1}{2\sqrt{u}}$, and $\dfrac{du}{dx} = -2x$.

Thus, $\dfrac{d}{dx}\left(25 - x^2\right)^{\frac{1}{2}} = \dfrac{d}{du}\left(25 - x^2\right)^{\frac{1}{2}} \times \dfrac{du}{dx} = \dfrac{-2x}{2\sqrt{u}} = \dfrac{-x}{\sqrt{25 - x^2}}$.

Then, by the **product rule**, we may deduce that:

$$f'(x) = 5\sqrt{25 - x^2} + \frac{-x}{\sqrt{25 - x^2}} \cdot 5x = 5\sqrt{25 - x^2} - \frac{5x^2}{\sqrt{25 - x^2}} = \frac{125 - 10x^2}{\sqrt{25 - x^2}}$$

Example (3.1-9). If $g(x) = \csc^3(2x + 1)$, find $g'(x)$.

Solution.

Let $u = \csc(2x + 1)$, then $y = g(x) = u^3$.

Since, $\dfrac{dy}{du} = 3u^2$, and $\dfrac{du}{dx} = -2\csc(2x + 1)\cot(2x + 1)$.

Thus, $\dfrac{dy}{dx} = \dfrac{dy}{du} \times \dfrac{du}{dx} = 3u^2 \cdot -2\csc(2x+1)\cot(2x+1) = -6\csc^3(2x+1)\cot(2x+1)$.

Example (3.1-10). If $y = \sin(\cos(2x))$, find $\dfrac{dy}{dx}$.

Solution.

Let $u = \cos(2x)$, then $y = \sin u$. Since, $\dfrac{dy}{du} = \cos u$, and $\dfrac{du}{dx} = -2\sin(2x)$.

Thus, $\dfrac{dy}{dx} = \dfrac{dy}{du} \times \dfrac{du}{dx} = \cos u \cdot -2\sin(2x) = -2\sin(2x)\cos(\cos(2x))$.

Example (3.2-1). Find $\dfrac{dy}{dx}$ if $x^2 + y^2 = 9$.

Solution.

Step 1: Differentiate each term of the equation with respect to x.

$$2x - 2y \cdot \dfrac{dy}{dx} = 0.$$

Step 2: Move all terms containing $\dfrac{dy}{dx}$ to the left side and all other terms to the right side.

$$-2y \cdot \dfrac{dy}{dx} = 2x.$$

Step 3: Factor out $\dfrac{dy}{dx}$ on the left side of the equation.

$$-y \cdot \dfrac{dy}{dx} = x.$$

Step 4: Solve for $\dfrac{dy}{dx}$.

$$\dfrac{dy}{dx} = -\dfrac{x}{y}.$$

This is actually perpendicular to the radius passing (x,y) whose slope is $\dfrac{y}{x}$.

Example (3.2-2). Find $\dfrac{dy}{dx}$ if $y^2 - 7y + x^2 - 4x = 10$.

Solution.

Step 1: Differentiate each term of the equation with respect to x.

$$2y\dfrac{dy}{dx} - 7\dfrac{dy}{dx} + 2x - 4 = 0.$$

Step 2: Move all terms containing $\dfrac{dy}{dx}$ to the left side and all other terms to the right side.

$$2y\frac{dy}{dx} - 7\frac{dy}{dx} = 4 - 2x.$$

Step 3: Factor out $\dfrac{dy}{dx}$ on the left hand side: $(2y - 7)\dfrac{dy}{dx} = 4 - 2x.$

Step 4: Solve for $\dfrac{dy}{dx}$: $\dfrac{dy}{dx} = \dfrac{4 - 2x}{2y - 7}.$

Example (3.2-3). Find $\dfrac{dy}{dx}$ if $x^3 + y^3 = 6xy$.

Solution.

Step 1: Differentiate both sides with respect to x: $3x^2 + 3y^2\,\dfrac{dy}{dx} = 6y + 6x\,\dfrac{dy}{dx}$

Step 2: Move all terms containing $\dfrac{dy}{dx}$ to the left side and all other terms to the right side.

$$3y^2\,\frac{dy}{dx} - 6x\,\frac{dy}{dx} = 6y - 3x^2$$

Step 3: Factor out $\dfrac{dy}{dx}$ on the left side: $(3y^2 - 6x)\dfrac{dy}{dx} = 6y - 3x^2.$

Step 4: Solve for $\dfrac{dy}{dx}$: $\dfrac{dy}{dx} = \dfrac{6y - 3x^2}{3y^2 - 6x} = \dfrac{2y - x^2}{y^2 - 2x}.$

Example (3.2-4). Find $\dfrac{\mathrm{d}y}{\mathrm{d}x}$ if $x\sin y = \cos(x+y)$.

Solution.

Step 1: Differentiate both sides with respect to x:

$$\sin y + x\cos y\,\frac{\mathrm{d}y}{\mathrm{d}x} = -\sin(x+y)\left(1 + \frac{\mathrm{d}y}{\mathrm{d}x}\right)$$

Step 2: Move all terms containing $\dfrac{\mathrm{d}y}{\mathrm{d}x}$ to the left side and all other terms to the right side.

$$x\cos y\,\frac{\mathrm{d}y}{\mathrm{d}x} + \sin(x+y)\,\frac{\mathrm{d}y}{\mathrm{d}x} = -\sin y - \sin(x+y)$$

Step 3: Factor out $\dfrac{\mathrm{d}y}{\mathrm{d}x}$ on the left side: $(x\cos y + \sin(x+y))\,\dfrac{\mathrm{d}y}{\mathrm{d}x} = -\sin y - \sin(x+y).$

Step 4: Solve for $\dfrac{\mathrm{d}y}{\mathrm{d}x}$: $\dfrac{\mathrm{d}y}{\mathrm{d}x} = -\dfrac{\sin y + \sin(x+y)}{x\cos y + \sin(x+y)}.$

Example (3.2-5). Where is the tangent to the curve $4x^2 + 9y^2 = 36$ vertical?

Solution.

Step 1: Differentiate both sides with respect to x: $8x + 18y\,\dfrac{\mathrm{d}y}{\mathrm{d}x} = 0$

Step 2: Move all terms containing $\dfrac{\mathrm{d}y}{\mathrm{d}x}$ to the left side: $18y\,\dfrac{\mathrm{d}y}{\mathrm{d}x} = -8x$

Step 3: Factor out $\dfrac{\mathrm{d}y}{\mathrm{d}x}$ on the left side: $9y\,\dfrac{\mathrm{d}y}{\mathrm{d}x} = -4x.$

Step 4: Solve for $\dfrac{\mathrm{d}y}{\mathrm{d}x}$: $\dfrac{\mathrm{d}y}{\mathrm{d}x} = -\dfrac{4x}{9y}.$

Since the tangent is vertical, $9y = 0 \Rightarrow y = 0$, in this case $x = \pm 3$

Example (3.3-1). For $f(x) = e^x$, show that $\dfrac{\mathrm{d}}{\mathrm{d}x}(\ln x) = \dfrac{1}{x}$

Solution.

Method 1: Let $g(x) = f^{-1}(x) = \ln x$. Since $f'(x) = e^x$, then $f'(g(x)) = e^{g(x)} = e^{\ln x} = x$, thus:

$$g'(x) = \frac{1}{f'(g(x))} = \frac{1}{x}.$$

Method 2: Let $y = \ln x$, then $x = e^y$. Differentiating with respect to y, we get:

$$\frac{\mathrm{d}x}{\mathrm{d}y} = e^y \quad \Rightarrow \quad \frac{\mathrm{d}y}{\mathrm{d}x} = \frac{1}{\mathrm{d}x/\mathrm{d}y} = \frac{1}{e^y} = \frac{1}{x}.$$

Method 3: The inverse of $y = e^x$ is $x = e^y$, differentiate implicitly with respect to x, we get:

$$1 = e^y\,\frac{\mathrm{d}y}{\mathrm{d}x} \quad \Rightarrow \quad \frac{\mathrm{d}y}{\mathrm{d}x} = \frac{1}{e^y} = \frac{1}{x}.$$

Example (3.3-2). Consider $f(x) = x^3 + 2x - 10$, if f^{-1} exists find $\left(f^{-1}\right)'(x)$

Solution.

Method 1:

Step 1: Let $y = f(x) = x^3 + 2x - 10$.

Step 2: Interchange x and y to obtain the expression for the inverse function

$$x = y^3 + 2y - 10.$$

Step 3: Differentiate with respect to y:

$$\frac{\mathrm{d}x}{\mathrm{d}y} = 3y^2 + 2.$$

Step 4: Take the reciprocal to get $\dfrac{\mathrm{d}y}{\mathrm{d}x} = \dfrac{1}{3y^2 + 2}$.

Method 2:

Step 1: Let $y = f(x) = x^3 + 2x - 10$.

Step 2: Interchange x and y to obtain the expression for the inverse function

$$x = y^3 + 2y - 10.$$

Step 3: Differentiate implicitly with respect to x:

$$1 = 3y^2 \frac{\mathrm{d}y}{\mathrm{d}x} + 2\frac{\mathrm{d}y}{\mathrm{d}x}.$$

Step 4: Solve for $\dfrac{\mathrm{d}y}{\mathrm{d}x}$:

$$(3y^2 + 2)\frac{\mathrm{d}y}{\mathrm{d}x} = 1 \quad \Rightarrow \quad \frac{\mathrm{d}y}{\mathrm{d}x} = \frac{1}{3y^2 + 2}$$

Example (3.3-3). For $f(x) = x^5 + 3x - 8$, find $\left(f^{-1}\right)'(-8)$

Solution.

By the previous COROLLARY, we may solve this problem using the following steps:

Step 1: Let $b = -8$, solve for a from $f(a) = b$.

$$f(a) = -8 \Rightarrow a^5 + 3a - 8 = -8 \Rightarrow a^5 + 3a = 0 \Rightarrow a(a^4 + 3) = 0 \Rightarrow a = 0$$

Step 2: Differentiate $f(x)$ and find $f'(a)$

$$f'(0) = 5x^4 + 3\big|_{x=0} = 3.$$

Step 3: Take the reciprocal to find $\left(f^{-1}(b)\right)'$:

$$\left(f^{-1}\right)'(-8) = \frac{1}{f'(0)} = \frac{1}{3}.$$

Example (3.3-4). For $f(x) = \cos x$ $(0 \le x \le \pi)$, find $\left(f^{-1}\right)'\left(\frac{1}{2}\right)$

Solution.

Step 1: Let $b = \frac{1}{2}$, solve for a from $f(a) = b$.

$$f(a) = \frac{1}{2} \Rightarrow \cos a = \frac{1}{2} \Rightarrow a = \frac{\pi}{3}.$$

Step 2: Differentiate $f(x)$ and find $f'(a)$

$$f'\left(\frac{\pi}{3}\right) = -\sin x\big|_{x=\frac{\pi}{3}} = -\frac{\sqrt{3}}{2}.$$

Step 3: Take the reciprocal to find $\left(f^{-1}(b)\right)'$:

$$\left(f^{-1}\right)'\left(\frac{1}{2}\right) = \frac{1}{f'\left(\frac{\pi}{3}\right)} = -\frac{2}{\sqrt{3}} = -\frac{2\sqrt{3}}{3}.$$

Example (3.4-1). Find $g'(x)$ if $g(x) = \arcsin x + \sqrt{1 - x^2}$

Solution. Since $g(x) = \arcsin x + \left(1 - x^2\right)^{\frac{1}{2}}$, by the previous formula and the chain rule:

$$g'(x) = \frac{1}{\sqrt{1 - x^2}} + \frac{1}{2} \cdot \left(1 - x^2\right)^{-\frac{1}{2}} \cdot (-2x)$$

$$= \frac{1}{\sqrt{1 - x^2}} - \frac{x}{\sqrt{1 - x^2}} = \frac{1 - x}{\sqrt{1 - x^2}}$$

Example (3.4-2). Find $f'(x)$ if $f(x) = \tan^{-1}\sqrt{x}$

Solution.

Using the chain rule, we let $u = \sqrt{x} = x^{\frac{1}{2}}$, then $y = f(x) = \tan^{-1} u$.
$\frac{dy}{du} = \frac{1}{1 + u^2} = \frac{1}{1 + x}$ and $\frac{du}{dx} = \frac{1}{2}x^{-\frac{1}{2}} = \frac{1}{2\sqrt{x}}$, thus:

$$f'(x) = \frac{dy}{dx} = \frac{dy}{du} \times \frac{du}{dx} = \frac{1}{2\sqrt{x}(1 + x)}.$$

Example (3.4-3). If $y = \cos^{-1}\left(\dfrac{1}{x}\right)$, find $\dfrac{dy}{dx}$

Solution.

Using the chain rule, we let $u = \dfrac{1}{x} = x^{-1}$, then $y = \cos^{-1} u$.

$\dfrac{dy}{du} = -\dfrac{1}{\sqrt{1-u^2}}$ and $\dfrac{du}{dx} = -x^{-2} = -\dfrac{1}{x^2}$, thus:

$$\frac{dy}{dx} = \frac{dy}{du} \times \frac{du}{dx} = -\frac{1}{\sqrt{1-u^2}} \cdot \left(-\frac{1}{x^2}\right)$$

$$= \frac{1}{\sqrt{1-u^2}} \cdot \frac{1}{x^2}$$

$$= \frac{1}{x^2\sqrt{1-\dfrac{1}{x^2}}} = \frac{1}{x^2 \cdot \dfrac{1}{|x|}\sqrt{x^2-1}}$$

$$= \frac{1}{|x|\sqrt{x^2-1}}.$$

Example (3.4-4). If $\arcsin x = \ln y$, find $\dfrac{dy}{dx}$

Solution. Using implicit differentiation, we differentiate both sides:

$$\frac{1}{\sqrt{1-x^2}} = \frac{1}{y} \cdot \frac{dy}{dx} \quad \Rightarrow \quad \frac{dy}{dx} = \frac{y}{\sqrt{1-x^2}}$$

Example (3.5-1). If $f(x) = \sqrt{x}$, find $f''(4)$

Solution. Rewrite $f(x) = \sqrt{x} = x^{\frac{1}{2}}$, then differentiate twice

$$f'(x) = \frac{1}{2}x^{-\frac{1}{2}}$$

$$f''(x) = -\frac{1}{2} \cdot \frac{1}{2}x^{-\frac{3}{2}} = -\frac{1}{4\sqrt{x^3}}$$

$$\text{Therefore, } f''(4) = -\frac{1}{4\sqrt{4^3}} = -\frac{1}{32}$$

Example (3.5-2). If $y = x\cos x$, find $\dfrac{d^2y}{dx^2}$

Solution. We need to use the **product rule** to differentiate the function:

$$\frac{dy}{dx} = \cos x - x\sin x$$

$$\frac{d^2y}{dx^2} = -\sin x - (\sin x + x\cos x) = -2\sin x - x\cos x$$

Example (3.5-3). If $x - y^2 = m$, find $\dfrac{\mathrm{d}^2 y}{\mathrm{d}x^2}$ at the point where $y = -1$

Solution. We need to use implicit differentiation with respect to x:

$$1 - 2y\,\frac{\mathrm{d}y}{\mathrm{d}x} = 0 \quad \Rightarrow \quad \frac{\mathrm{d}y}{\mathrm{d}x} = \frac{1}{2y} = \frac{1}{2}y^{-1}$$

$$\frac{\mathrm{d}^2 y}{\mathrm{d}x^2} = \frac{\mathrm{d}}{\mathrm{d}x}\left(\frac{\mathrm{d}y}{\mathrm{d}x}\right) = \frac{\mathrm{d}}{\mathrm{d}x}\left(\frac{1}{2}y^{-1}\right) = -\frac{1}{2}y^{-2}\cdot\frac{\mathrm{d}y}{\mathrm{d}x} = -\frac{1}{2}y^{-2}\cdot\frac{1}{2}y^{-1} = -\frac{1}{4y^3}$$

when $y = -1$, $\dfrac{\mathrm{d}^2 y}{\mathrm{d}x^2} = -\dfrac{1}{4(-1)^3} = \dfrac{1}{4}$

11.4   Unit 4 Solutions

Example (4.1-1). Let $G = 400(15 - t)^2$ be the number of gallons of water in a cistern t minutes after an outlet pipe is opened. Find the average rate of drainage during the first 5 minutes and the instantaneous rate at which the water is running out at the end of 5 minutes, and explain these rates in context.

Solution.

The average rate of change during the first 5 minutes is:

$$\frac{G(5) - G(0)}{5 - 0} = \frac{400 \cdot 10^2 - 400 \cdot 15^2}{5} = -10,000 \text{ gal/min.}$$

The instantaneous rate of change at $t = 5$ is $G'(5)$, which is given by:

$$G'(t) = (-1) \cdot 2 \cdot 400(15 - t) = -800(15 - t)$$
$$G'(5) = -800 \cdot (15 - 5) = -8,000 \text{ gal/min.}$$

The gallons of water in a cistern is **decreasing** at an average rate of -10,000 gal/min in the first 5 minutes, and it is **decreasing** at a rate of -8,000 gal/min at $t = 5$ minutes.

Example (4.1-2). The function $t = f(M)$ models the time, in seconds, for a chemical reaction to occur as the mass M of a catalyst used, measured in grams. What is the unit for $f'(M)$

Solution. The unit is: seconds per grams (s/g)

Example (4.2-1). The position of a particle moving on a straight line is $x(t) = 2t^3 - 10t^2 + 5$. Find:

(a) the position at $t = 1$

(b) the instantaneous velocity at $t = 1$

(c) the acceleration at $t = 1$

(d) the speed of the particle at $t = 1$.

Solution.

(a) $x(1) = 2(1)^3 - 10(1)^2 + 5 = -3$

(b) $v(t) = x'(t) = 6t^2 - 20t \Rightarrow v(1) = 6(1)^2 - 20(1) = -14$

(c) $a(t) = v'(t) = 12t - 20 \Rightarrow a(1) = 12(1) - 20 = -8$

(d) speed $= |v(1)| = |-14| = 14$

Example (4.2-2). A particle moves along the x-axis so that its position is $x(t) = -t - e^{1-t}$.

(a) Find the velocity function.

(b) Find the acceleration function.

(c) Is the speed of the particle increasing at time $t = 4$?

(d) Find all values of t at which the particle changes direction.

(e) Find the total distance traveled by the particle over the time interval $0 \leq t \leq 4$.

Solution.

(a) $v(t) = s'(t) = -1 + e^{1-t}$

(b) $a(t) = v'(t) = -e^{1-t}$

(c) $v(4) = -1 + e^{-3} = -1 + \dfrac{1}{e^3} < 0$ and $a(4) = -e^{-3} < 0$

　　The velocity and acceleration have the same sign, thus the speed is increasing.

(d) Let $v(t) = 0 \Rightarrow -1 + e^{1-t} = 0 \Rightarrow e^{1-t} = 1 \Rightarrow t = 1$

　　When $t < 1$, $1 - t > 0 \Rightarrow v(t) > 0$, when $t > 1$, $1 - t < 0 \Rightarrow v(t) < 0$. The particle changes direction at $t = 1$.

(e) Since $v(t) > 0$ for $0 \leq t < 1$ and $v(t) < 0$ for $1 < t \leq 4$,

　　Total Distance $= |x(1) - x(0)| + |x(4) - x(1)| = |-2 + e| + \left|-2 - (-4 - e^{-3})\right| = e + e^{-2}$

Example (4.2-3). A particle moves along the x-axis so that its position is $x(t) = t^3 - 6t^2 + 9t + 3$.

(a) For $0 \leq t \leq 6$, find all times t during the particle is moving to the right.

(b) Find the acceleration at time $t = 4$. Is the speed of the particle increasing at $t = 4$?

(c) Find all times t in the open interval $0 < t < 6$ when particle changes direction.

(d) Find the total distance traveled by the particle from $t = 0$ until time $t = 6$.

(e) During $0 \leq t \leq 4$, what is the greatest distance between the particle from the origin?

Solution.

(a) $v(t) = 3t^2 - 12t + 9 > 0 \Rightarrow 3(t-1)(t-3) > 0$

　　Thus, the particle is moving to the right when $0 < t < 1$ and $3 < t < 6$

(b) $a(t) = v'(t) = 6t - 12$, $v(4) = 3(4)^2 - 12(4) + 9 = 9$ and $a(4) = 6(4) - 12 = 12$

　　Since $v(t)$, $a(t)$ have the same sign, the speed is increasing.

(c) $v(4) = -1 + e^{-3} = -1 + \dfrac{1}{e^3} < 0$ and $a(4) = -e^{-3} < 0$

　　The velocity and acceleration have the same sign, thus the speed is increasing.

(d) Let $v(t) = 0 \Rightarrow 3t^2 - 12t + 9 = 0 \Rightarrow 3(t-1)(t-3) = 0 \Rightarrow t = 1$ and $t = 3$

　　Distance $= |x(1) - x(0)| + |x(3) - x(1)| + |x(6) - x(3)| = |7 - 3| + |3 - 7| + |57 - 3| = 62$

(e) $x(1) = 7$, $x(3) = 3$, $x(4) = 7$. The greatest distance from the origin is 7 .

Example (4.2-4).

The graph of the velocity is show in the figure.

  (a) When is the acceleration $= 0$?

  (b) When is the particle moving to the right?

  (c) When is the speed the greatest?

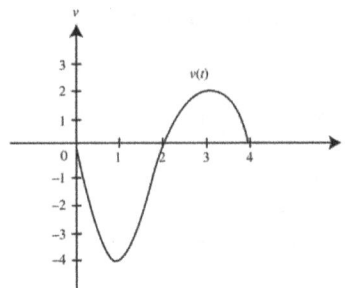

Solution.

  (a) $a(t) = v'(t)$, which is the slope of the graph.

     At $t = 1$ and $t = 3$, the graph has horizontal tangent (slope $= 0$).

  (b) For $2 < t < 4$, $v(t) > 0$. Thus the particle is moving to the right during $2 < t < 4$.

  (c) Speed $= |v(t)|$, at $t = 1$, the graph has greatest absolute value.

     Therefore, the speed at $t = 1$ is the greatest.

Example (4.2-5). The graph of the position function of a moving particle is shown in the rogue below.

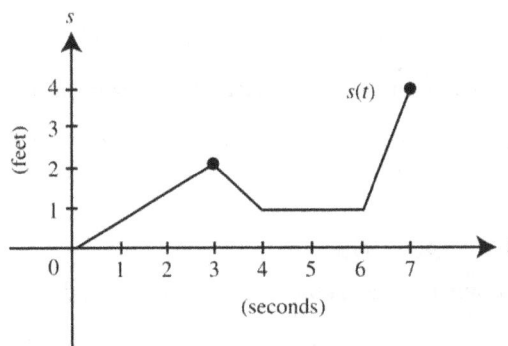

  (a) What is the particle's position at $t = 5$?

  (b) When is the particle moving to the left?

  (c) When is the particle standing still?

  (d) When does the particle have the greatest speed?

Solution.

  (a) At $t = 5$, $s(5) = 1$.

  (b) For $3 < t < 4$, $s(t)$ decreases. Thus, the particle moves to the left when $3 < t < 4$.

  (c) When $4 < t < 6$, the particles stays at 1.

  (d) When $6 < t < 7$, the slope is greatest, thus the speed is greatest.

Example (4.3-1). If x and y are both differentiable functions of t and are related by the equation $y = x^2 + 5$ and $\dfrac{dx}{dt} = 3$ when $x = 2$. Find $\dfrac{dy}{dy}$ when $x = 2$.

Solution.

The relationship between the two rates are given by

$$\frac{dy}{dt} = \frac{dy}{dx} \cdot \frac{dx}{dt}$$

When $x = 2$, the rates are: $\left.\frac{dy}{dx}\right|_{x=2} = 2x|_{x=2} = 4$ and $\frac{dx}{dt} = 3$.

Solving the equation linking rates, we get:

$$\frac{dy}{dt} = \frac{dy}{dx} \cdot \frac{dx}{dt} = 4 \times 3 = 12.$$

Example (4.3-2). Find the surface area of a sphere at the instant when the rate of increase of the volume of the sphere is nine times the rate of increase of the radius.

Solution.

The relationship between the two rates are given by

$$\frac{dV}{dt} = \frac{dV}{dr} \cdot \frac{dr}{dt} = 9 \cdot \frac{dr}{dt} \quad \Rightarrow \quad \frac{dV}{dr} = 9$$

Since $V = \frac{4}{3}\pi r^3$, $\frac{dV}{dr} = 4\pi r^2$, which is also the formula for the surface area

By $\frac{dV}{dr} = 4\pi r^2 = 9$, the surface area is 9 square units.

Example (4.3-3). A water tank has the shape of a cylinder with radius 5 meters. Let h be the depth of the water in the tank, measured in meters, where h is a function of time, t, measured in seconds. The volume V of the water tank is changing at the rate of -15π cubic meters per second. Find $\frac{dh}{dt}$.

Solution.

The relationship between the two rates are given by

$$\frac{dV}{dt} = \frac{dV}{dh} \cdot \frac{dh}{dt}$$

Since $V = \pi r^2 h$, then $\frac{dV}{dh} = \pi r^2 = 25\pi$.

Solving the equation linking rates, we get:

$$\frac{dV}{dt} = \frac{dV}{dh} \cdot \frac{dh}{dt} \quad \Rightarrow \quad -15\pi = 25\pi \cdot \frac{dh}{dt}$$
$$\frac{dh}{dt} = -\frac{3}{5}\text{m/s}$$

Example (4.3-4). Suppose that liquid is to be cleared of sediment by allowing it to drain through a conical filter that is 12 cm high and has a radius of 6 cm at the top. If the liquid is forced out of the cone at a constant rate of 2 cm^3/min, how fast is the depth of the liquid decreasing at the instant when the liquid in the cone is 4 cm deep?

Solution.

The relationship between the two rates are given by

$$\frac{dV}{dt} = \frac{dV}{dh} \cdot \frac{dh}{dt}$$

Since $V = \frac{1}{3}\pi r^2 h$, and by the similarities of triangles, we get $r = h/2$, thus the relationship can be noted as:

$$V = \frac{\pi}{3} \cdot \frac{h^2}{4} \cdot h = \frac{\pi}{12}h^3 \Rightarrow \frac{dV}{dh} = \frac{\pi}{4}h^2$$

At $h = 4$, $\frac{dV}{dh} = 4\pi$, and $\frac{dV}{dt} = -2$

Solving the equation linking rates, we get:

$$\frac{dV}{dt} = \frac{dV}{dh} \cdot \frac{dh}{dt} \Rightarrow -2 = 4\pi \cdot \frac{dh}{dt} \Rightarrow$$

$$\frac{dh}{dt} = -\frac{1}{2\pi} \text{ cm/min}$$

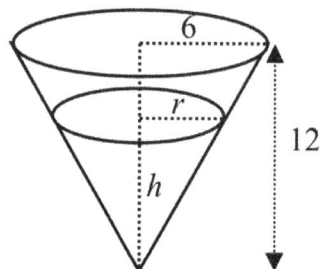

Example (4.3-5). A 25 feet- long ladder is leaning against the wall of a house and sliding away from the wall at a rate of 3 feet per second. How fast is the top of the ladder moving down the wall when the base of the ladder is 15 feet?

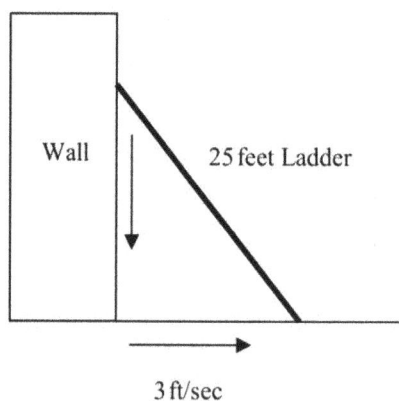

Solution.

The relationship between the two rates are given by

$$\frac{dy}{dt} = \frac{dy}{dx} \cdot \frac{dx}{dt}$$

We already know that $\frac{dx}{dt} = 3$ ft/s, and $x^2 + y^2 = 25$, differentiate both side with respect to x, we have

$$2x + 2y\frac{dy}{dx} = 0 \Rightarrow \frac{dy}{dx} = -\frac{x}{y}$$

When $x = 15$, $y = \sqrt{25^2 - 15^2} = 20$, then $\dfrac{dy}{dx} = -\dfrac{3}{4}$. Solving the equation linking rates, we get:

$$\frac{dy}{dt} = \frac{dy}{dx} \cdot \frac{dx}{dt} = -\frac{3}{4} \cdot 3 = -\frac{9}{4}\text{ft/s}.$$

The top of the ladder is moving down the wall with a speed of $-\dfrac{9}{4}$ feet per second.

Example (4.4-1). Write an equation of the tangent line to $f(x) = x^3$ at $(2, 8)$. Use the tangent line to find the approximate values of $f(1.9)$ and $f(2.01)$.

Solution.

Method 1: Using the equation of tangent Differentiate
$f(x)$: $f'(x) = 3x^2 \Rightarrow f'(2) = 3(2)^2 = 12$.

The equation of the tangent at $x = 2$ is:

$$y = f'(2)(x - 2) + f(2) = 12(x - 2) + 8 = 12x - 16$$

Then we can approximate the values:

$$f(1.9) \approx 12 \times 1.9 - 16 = 6.8$$
$$f(2.01) \approx 12 \times 2.01 - 16 = 8.12$$

Method 2: Directly use the approximation formula

$$f(1.9) \approx f(2) + f'(2)(1.9 - 2) = 8 - 12 \times 0.1 = 6.8$$
$$f(2.01) \approx f(2) + f'(2)(2.01 - 2) = 8 + 12 \times 0.01 = 8.12$$

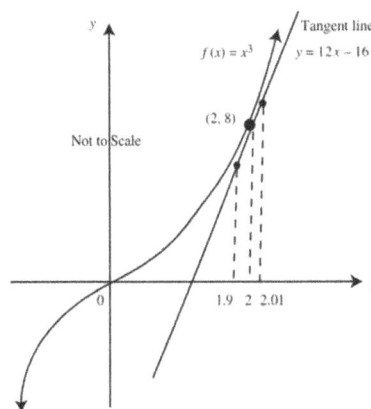

Example (4.4-2). Estimate the value of $\dfrac{3}{(1 - x)^2}$ at $x = 0.05$ from $x = 0$.

Solution.

The derivative of the function is: $f'(x) = \dfrac{6}{(1 - x)^3}$, thus $f(0) = 3$ and $f'(0) = 6$, thus:

$$f(0.05) \approx f(0) + f'(0)(0.05 - 0) = 3 + 6 \times 0.05 = 3.3$$

Example (4.4-3). Let $g(x) = e^x + 1$, and $g(0) = 2$ by considering the graph of function, find an approximation for $g(0.015)$ and state whether it is an overestimate or underestimate.

Solution.

The derivative of the function is: $g'(x) = e^x$, thus $g'(0) = 1$, thus:

$$g(0.015) \approx g(0) + g'(0)(0.015 - 0) = 2 + 1 \times 0.015 = 2.015$$

Since the graph of $g(x)$ is concave up, then the approximation is an underestimate
** In fact, $g(0.015) = e^{0.015} + 1 \approx 2.015113$

Example (4.5-1). Show that $\lim\limits_{x \to 0} \dfrac{\sin x}{x} = 1$

Solution. The indeterminate form is $\dfrac{0}{0}$, thus the limit is given by:

$$\lim_{x \to 0} \frac{\sin x}{x} = \lim_{x \to 0} \frac{\cos x}{1} = 1.$$

Example (4.5-2). Find $\lim\limits_{x \to 0} \dfrac{1 - \cos x}{x^2}$, if it exists

Solution. The indeterminate form is $\dfrac{0}{0}$, thus the limit is given by:

$$\lim_{x \to 0} \frac{1 - \cos x}{x^2} = \lim_{x \to 0} \frac{\sin x}{2x} = \lim_{x \to 0} \frac{\cos x}{2} = \frac{1}{2}.$$

Example (4.5-3). Find $\lim\limits_{x \to 0^+} \dfrac{\ln(x+1)}{\sqrt{x}}$, if it exists

Solution. The indeterminate form is $\dfrac{0}{0}$, thus the limit is given by:

$$\lim_{x \to 0^+} \frac{\ln(x+1)}{\sqrt{x}} = \lim_{x \to 0^+} \frac{\dfrac{1}{x+1}}{\dfrac{1}{2\sqrt{x}}} = \lim_{x \to 0^+} \frac{2\sqrt{x}}{x+1} = 0.$$

Example (4.5-4). Find $\lim\limits_{x\to 0}\dfrac{e^x-1}{\tan 2x}$, if it exists

Solution. The indeterminate form is $\dfrac{0}{0}$, thus the limit is given by:

$$\lim_{x\to 0}\frac{e^x-1}{\tan 2x}=\lim_{x\to 0}\frac{e^x}{2\sec^2 2x}=\frac{1}{2}.$$

Example (4.5-5). Find $\lim\limits_{x\to 0}\dfrac{2x^2}{e^{2x}-1}$, if it exists

Solution. The indeterminate form is $\dfrac{\infty}{\infty}$, thus the limit is given by:

$$\lim_{x\to 0}\frac{2x^2}{e^{2x}-1}=\lim_{x\to 0}\frac{4x}{2e^{2x}}=\lim_{x\to 0}\frac{4}{4e^{2x}}=1.$$

Example (4.5-6). Find $\lim\limits_{x\to\infty} x\cdot\sin\dfrac{1}{x}$, if it exists

Solution. The indeterminate form is $0\cdot\infty$, thus the limit is given by:

$$\lim_{x\to\infty}x\cdot\sin\frac{1}{x}=\lim_{x\to\infty}\frac{\sin\frac{1}{x}}{\frac{1}{x}}=\lim_{x\to\infty}\frac{-\frac{1}{x^2}\cdot\sin\frac{1}{x}}{-\frac{1}{x^2}}=\lim_{x\to\infty}\cos\frac{1}{x}=1$$

(Note that the limit is equivalent to $\lim\limits_{x\to 0}\dfrac{\sin x}{x}$)

Example (4.5-7). Find $\lim\limits_{x\to-\infty} xe^x$, if it exists

Solution. The indeterminate form is $0\cdot-\infty$, thus the limit is given by:

$$\lim_{x\to-\infty}xe^x=\lim_{x\to\infty}\frac{x}{e^{-x}}=\lim_{x\to\infty}\frac{1}{-e^{-x}}=-\lim_{x\to\infty}e^x=0$$

(Note that the limit is equivalent to $\lim\limits_{x\to 0^-}\dfrac{e^{\frac{1}{x}}}{x}$, which is an $\dfrac{0}{0}$ form)

Example (4.5-8). Find $\lim\limits_{x\to\infty} x^{\frac{1}{x}}$, if it exists

Solution. The indeterminate form is ∞^0, let the limit be equal to y and take the natural log:

$$\ln y=\lim_{x\to\infty}\frac{\ln x}{x}=\lim_{x\to\infty}\frac{\frac{1}{x}}{1}=\lim_{x\to\infty}\frac{1}{x}=0$$

Since $\ln y=0$, then $y=1$. Therefore, $\lim\limits_{x\to\infty}x^{\frac{1}{x}}=1$

199

Example (4.5-9). Find $\lim\limits_{x \to \infty} \left(1 + \dfrac{1}{x}\right)^x$, if it exists

Solution. The indeterminate form is 1^∞, let the limit be equal to y and take the natural log:

$$\ln y = \lim_{x \to \infty} x \cdot \ln\left(1 + \frac{1}{x}\right)$$

Here the indeterminate form changes to $0 \cdot \infty$, then we have:

$$\ln y = \lim_{x \to \infty} \frac{\ln\left(1 + \dfrac{1}{x}\right)}{\dfrac{1}{x}} = \lim_{x \to \infty} \frac{\dfrac{x}{x+1} \cdot \left(-\dfrac{1}{x^2}\right)}{-\dfrac{1}{x^2}} = \lim_{x \to \infty} \frac{x}{x+1} = 1$$

Since $\ln y = 1$, then $y = e$. Therefore, $\lim\limits_{x \to \infty} \left(1 + \dfrac{1}{x}\right)^x = e$

(Note that the limit is another important basic limit that is closely related to the definition of the number e.)

11.5   Unit 5 Solutions

Example (5.1-1). The function $f(x) = \dfrac{1}{4}x^3 + 1$ satisfies the **Mean Value Theorem** over the interval $[0,2]$. Find all values of c in the interval $(0,2)$ at which the tangent line to the graph of f is parallel to the secant line joining the points $(0, f(0))$ and $(2, f(2))$.

Solution.

When $x = 0,\ 2$, we have $f(0) = 1$, $f(2) = 3$, the slope of the secant line is:

$$\frac{f(2) - f(0)}{2 - 0} = \frac{3 - 1}{2 - 0} = 1$$

The derivative of the function $f(x)$ is given by:

$$f'(x) = \frac{3}{4}x^2 \ \Rightarrow \ f'(c) = \frac{3}{4}c^2$$

Since the tangent is parallel to the secant,

$$f'(c) = \frac{3}{4}c^2 = 1 \ \Rightarrow \ 3c^2 = 4 \ \Rightarrow \ c = \frac{2}{\sqrt{3}} = \frac{2\sqrt{3}}{3} \text{ Since } c \text{ lies in the interval } (0,2)$$

Example (5.1-2). (Calculator required) The function f is defined by $f(x) = 3x - 4\cos{(2x + 1)}$. What are all values of x that satisfy the conclusion of the Mean Value Theorem applied to f on the interval $[-1, 2]$?

Solution.

When $x = -1,\ 2$, we have $f(1) = 6 - 4\cos{(5)}$, $f(2) = -3 - 4\cos{(-1)}$, the average rate of change is:

$$\frac{f(2) - f(-1)}{2 - (-1))} = \frac{9 - 4\cos{(5)} + 4\cos{(-1)}}{3} = 3.34129$$

The derivative of the function $f(x)$ is given by: $f'(x) = 3 + 8\sin{(2x + 1)}$ By the MVT,

$$f'(x) = 3 + 8\sin{(2x + 1)} = 3.34219 \ \Rightarrow \ x = -0.479 \text{ and } 1.049 \text{ on the interval } [-1, 2].$$

Example (5.1-3). Verify that the function $f(x) = 2x^2 - 8x + 6$ defined over the interval $[1, 3]$ satisfies the conditions of Rolle's theorem. Find all points c guaranteed by Rolle's theorem.

Solution.

When $x = 1,\ 3$, we have $f(1) = f(3) = 0$, thus the Rolle's Theorem applies.

The derivative of the function is given by

$$f'(x) = 4x - 8$$

By the Rolle's Theorem,

$$f'(x) = 4x - 8 = 0 \ \Rightarrow \ x = 2.$$

Example (5.2-1). Determine all critical points for $f(x) = \sqrt[3]{x^2}(2x + 1)$ within $\left[-\dfrac{1}{2}, \dfrac{1}{2}\right]$.

Solution.

Since $f(x) = \sqrt[3]{x^2}(2x+1) = x^{\frac{2}{3}}(2x+1) = 2x^{\frac{5}{3}} + x^{\frac{2}{3}}$

The derivative is given by:

$$f'(x) = \frac{10}{3}x^{\frac{2}{3}} + \frac{2}{3}x^{-\frac{1}{3}} = \frac{10x^{\frac{2}{3}}}{3} + \frac{2}{3x^{\frac{1}{3}}}$$

$$= \frac{10x^{\frac{2}{3}} \cdot x^{\frac{1}{3}}}{3x^{\frac{1}{3}}} + \frac{2}{3x^{\frac{1}{3}}}$$

$$= \frac{10x+2}{3x^{\frac{1}{3}}} = \frac{10x+2}{3\cdot\sqrt[3]{x}}$$

If $f'(x) = 0 \Rightarrow x = -\frac{1}{5}$.   If $f'(x)$ DNE $\Rightarrow x = 0$

Therefore, there are two critical points for this function:

$$x = -\frac{1}{5} \text{ and } x = 0$$

The graph of the function is shown on the right.

(Note that the function f is not differentiable at $x = 0$ since there is a sharp corner at this point)

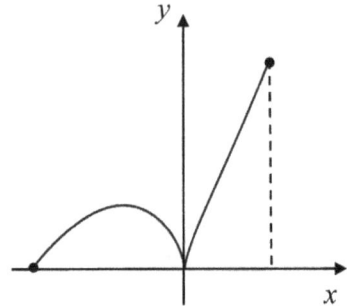

Example (5.2-2). Determine all critical points and stationary points for $h(t) = 10t \cdot e^{3-t^2}$.

Solution.

The derivative of $h(t)$ is given by:

$$h'(t) = 10e^{3-t^2} + 10te^{3-t^2}(-2t) = 10e^{3-t^2} - 20t^2e^{3-t^2} = 10e^{3-t^2}\left(1 - 2t^2\right)$$

We may find that $h'(t)$ is defined for all real values of t, thus all critical points are stationary points. Then, let $h'(t) = 0$, we get

$$h'(t) = 10e^{3-t^2}\left(1 - 2t^2\right) = 0 \Rightarrow 1 - 2t^2 = 0 \Rightarrow t^2 = \frac{1}{2} \Rightarrow t = \pm\frac{\sqrt{2}}{2}$$

Therefore, there are two critical points for this function:

$$t = \pm\frac{\sqrt{2}}{2}$$

Both of them are stationary points.

Example (5.3-1). Find the intervals on which $f(x) = x^3 - 3x^2$ is increasing or decreasing.

Solution.

$f'(x) = 3x^2 - 6x = 0 \Rightarrow 3x(x-2) = 0.$ Critical points: $x = 0,\ 2$

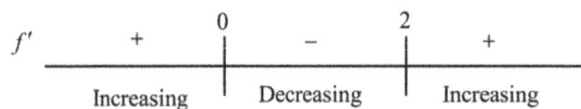

$$
\begin{array}{ccccccc}
 & & 0 & & 2 & \\
f' & \underline{\quad + \quad} & | & \underline{\quad - \quad} & | & \underline{\quad + \quad} \\
 & \text{Increasing} & & \text{Decreasing} & & \text{Increasing}
\end{array}
$$

When $-\infty < x < 0$, $f'(x) > 0 \Rightarrow$ increasing on $(-\infty, 0]$

When $0 < x < 2$, $f'(x) < 0 \Rightarrow$ decreasing on $[0, 2]$

When $2 < x < \infty$, $f'(x) > 0 \Rightarrow$ increasing on $[2, \infty)$

Example (5.3-2). Find the intervals on which $f(x) = \dfrac{x^2}{x+1}$ is increasing or decreasing.

Solution.

The derivative of $f(x)$ is

$$f'(x) = \frac{(x+1)(2x) - x^2}{(x+1)^2} = \frac{x^2 + 2x}{(x+1)^2} = \frac{x(x+2)}{(x+1)^2}$$

Critical points at $x = 0, 2$.   Asymptote at $x = -1$

Note: $x = 1$ is not the critical point, because it is not in the domain of f.

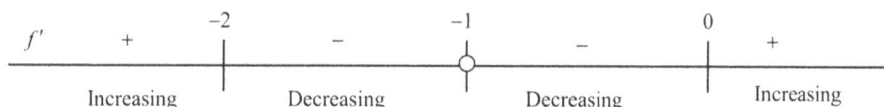

The function $f(x)$ is increasing on $(-\infty, 2]$ and $[0, \infty]$, and decreasing on $[-2, -1]$ and $[-1, 0]$

Example (5.4-1). Find the relative extrema for the function $f(x) = \dfrac{x^3}{3} - x^2 - 3x$.

Solution.

The derivative of $f(x)$ is: $f'(x) = x^2 - 2x - 3$, which is defined for all x.

Thus, we can find the critical points (stationary points) by setting $f'(x) = 0$, then:

$$f'(x) = x^2 - 2x - 3 = 0$$
$$(x - 3)(x + 1) = 0$$
$$x = 3 \text{ or } x = -1.$$

$$f'(x) = x^2 - 2x - 3 = 0 \Rightarrow (x - 3)(x + 1) = 0 \Rightarrow x = 3 \text{ or } x = -1.$$

The second derivative is given by: $f''(x) = 2x - 2$, then:

$$f''(3) = 2(3) - 2 = 4 \Rightarrow f(3) \text{ is a relative minimum.}$$
$$f''(-1) = 2(-1) - 2 = -4 \Rightarrow f(-1) \text{ is a relative maximum.}$$

Since $f(3) = \dfrac{3^3}{3} - 3^2 - 3 \cdot 3 = -9$ and $f(-1) = \dfrac{(-1)^3}{3} - (-1)^2 - 3 \cdot (-1) = \dfrac{5}{3}$

Therefore, -9 is a relative minimum of f and $\dfrac{5}{3}$ is a relative maximum of f.

Example (5.4-2). Find the relative maximum or relative minimum of $f(x) = \left(x^2 - 1\right)^{\frac{2}{3}}$.

Solution.

The derivative of $f(x)$ is

$$f'(x) = \frac{2}{3}\left(x^2 - 1\right)^{-\frac{1}{3}} \cdot (2x)$$
$$= \frac{4}{3}x\left(x^2 - 1\right)^{-\frac{1}{3}}$$
$$= \frac{4x}{3 \cdot \sqrt[3]{x^2 - 1}}$$

203

When $x = \pm 1$, $f(x)$ is defined but $f'(x)$ does not exist, thus $x = \pm 1$ are critical points.

Setting $f'(x) = 0 \Rightarrow x = 0$, this is a stationary point.

The sign of f' is shown in the following figure

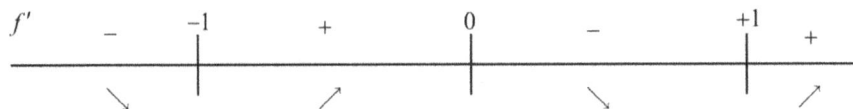

From the graph above, we may notice that:

1. f' changes from positive to negative at $x = 0$.

2. f' changes from negative to positive at $x = \pm 1$.

When $x = 0$, function f has relative maximum $f(0) = 1$

When $x = \pm 1$, function f has relative minimum $f(-1) = 0$, $f(1) = 0$

Example (5.4-3). Find the relative extrema of $f(x) = \dfrac{x^4 + 1}{x^2}$.

Solution.

Since $f(x) = \dfrac{x^4 + 1}{x^2} = x^2 + x^{-2}$. The derivative of $f(x)$ is

$$f'(x) = 2x - 2x^{-3} = x^{-3}\left(2x^4 - 2\right) = \frac{2\left(x^4 - 1\right)}{x^3} = \frac{2\left(x^2 + 1\right)(x+1)(x-1)}{x^3}$$

(You can also use the *quotient rule* to find $f'(x)$)

When $x = 0$, $f(x)$ and $f'(x)$ are both not defined thus $x = 0$ is an asymptote, not a critical point.

Setting $f'(x) = 0 \Rightarrow x = \pm 1$, these are stationary points.

The sign of f' is shown in the following figure

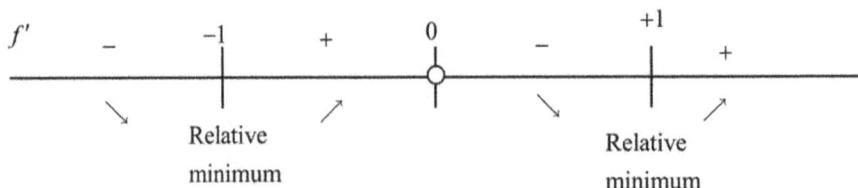

When $x = \pm 1$, function f has relative minimum $f(-1) = 2$, $f(1) = 2$

Since f is not continuous at $x = 0$, thus the function has no maximum.

Example (5.4-4). Show that the absolute minimum of $f(x) = \sqrt{25 - x^2}$ on $[-5, 5]$ is 0 and the absolute maximum is 5 .

Solution.

Since $f(x) = \left(25 - x^2\right)^{\frac{1}{2}}$. The derivative of $f(x)$ is

$$f'(x) = \frac{1}{2}\left(25 - x^2\right)^{-\frac{1}{2}} \cdot (-2x) = -\frac{2x}{2\sqrt{25 - x^2}} = -\frac{x}{\sqrt{25 - x^2}}$$

When $x = \pm 5$, $f(x)$ is defined and $f'(x)$ does not exit, thus $x = \pm 5$ are critical points.

Setting $f'(x) = 0 \Rightarrow x = 0$, these are stationary points.

Since The second derivative of is given by:

$$f''(x) = \frac{-\sqrt{25 - x^2} - (-x) \cdot -\dfrac{x}{\sqrt{25 - x^2}}}{25 - x^2} = -\frac{1}{\sqrt{25 - x^2}} - \frac{x^2}{(25 - x^2)^{\frac{3}{2}}}$$

When $x = 0$, $f''(0) = -\dfrac{1}{5}$, thus $f(0) = 5$ is a relative maximum

Now we check the endpoints, $f(-5) = 0$, $f(5) = 0$.

Thus, 0 is the absolute minimum value, 5 is the absolute maximum value.

Example (5.5-1). Find the points of inflection of $f(x) = (x - 1)^{\frac{1}{3}}$ and determine the intervals where the function f is concave up and where the function is concave down.

Solution.

The first and second derivatives of $f(x)$ is

$$f'(x) = \frac{1}{3}(x - 1)^{-\frac{2}{3}} \qquad f''(x) = -\frac{2}{9}(x - 1)^{-\frac{5}{3}} = -\frac{2}{9 \cdot \sqrt[3]{(x - 1)^5}}$$

$f''(x)$ is always non-zero, but When $x = 1$, $f(x)$ is defined but $f''(x)$ does not exit.

Since $\sqrt[3]{(x - 1)^5} < 0$ when $x < 1$ and $\sqrt[3]{(x - 1)^5} > 0$ when $x > 1$

$f''(x) > 0$ when $x < 1$ and $f''(x) < 0$ when $x > 1$

Since $f''(x)$ changes sign at $x = 1$

Thus, $f(x)$ is concave up on $(-\infty, 1)$ and concave down on $(1, \infty)$

By $f(1) = 0$, the point $(1, 0)$ is the point of inflection.

Example (5.5-2). Given $f(x) = x + \sin x$, $(0 \le x \le 2\pi)$, find all points of inflection of f.

Solution.

The first and second derivatives of $f(x)$ is

$$f'(x) = 1 + \cos x \qquad f''(x) = -\sin x$$

Set $f''(x) = 0 \Rightarrow x = 0$, π, 2π, where $f''(x)$ changes sign only at $x = \pi$

By $f(\pi) = \pi$, then (π, π) is a point of inflection.

(Note that the *endpoints* cannot be recognized as the point of inflection, since the concavity at one side is unknown.)

Example (5.5-3). The graph of f is shown in the figure below and f is twice differentiable. Which of the following statements is true?

(A) $f(5) < f'(5) < f''(5)$

(B) $f''(5) < f'(5) < f(5)$

(C) $f'(5) < f(5) < f''(5)$

(D) $f'(5) < f''(5) < f(5)$

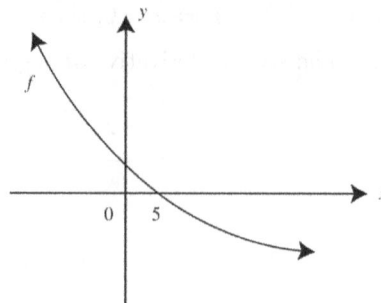

Solution.

The graph indicates that $f(5) = 0$, $f'(5) < 0$, since f is decreasing

and $f''(5) > 0$, since f is concave upward.

Thus, $f'(5) < f(5) < f''(5)$, choice (C).

Example (5.5-4).

The graph of f is shown in the figure below. Find the points of inflection of f and determine where the function f is concave upwards and where it is concave downwards on $[-3, 5]$.

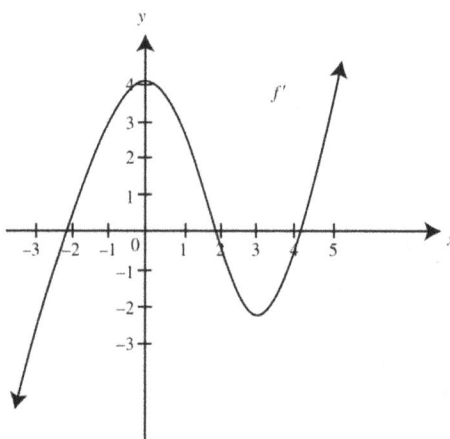

Solution.

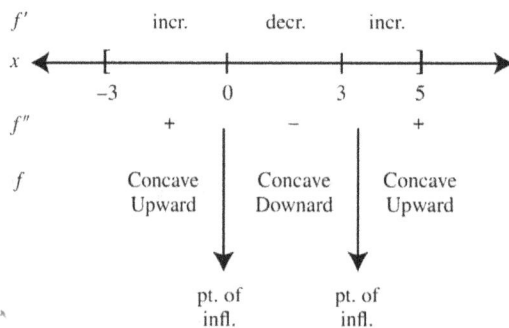

f is concave upwards on $[-3, 0)$ and $(3, 5]$, and is concave downwards on $(0, 3)$. There are two points of inflection: one at $x = 0$ and the other at $x = 3$.

Example (5.6-1).

Sketch the graph of $f(x) = x^3 - 3x^2 - 24x + 32$.

Solution.

Use the first and second derivative test:

$$f'(x) = 3x^2 - 6x - 24 = 0 \Rightarrow 3(x-4)(x+2) = 0 \Rightarrow x = 4 \text{ or } x = -2$$

$$f''(x) = 6x - 6 = 0 \Rightarrow x = 1$$

The function f have the following features:

1. Relative max and min: $f(-2) = 60$ and $f(4) = -48$

2. Inflection point: $f(1) = 6$

3. y-intercept: $f(0) = 32$

The graph of the function is given by:

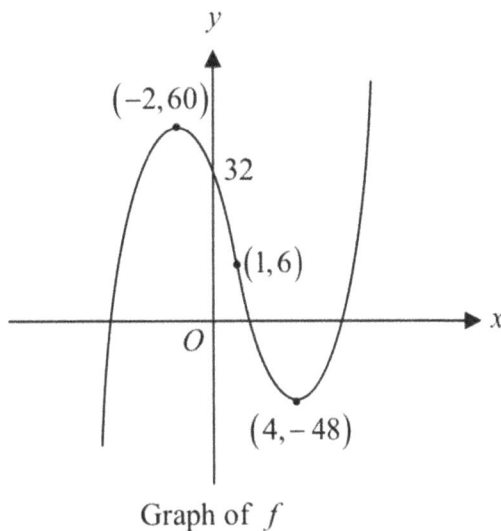

Graph of f

Example (5.6-2).

Sketch the graph of $f(x) = \dfrac{x^2 - 3x + 2}{x}$.

Solution.

Since $f(x) = x - 3 + \dfrac{2}{x}$, then the asymptotes of f are:

$$x = 0 \text{ (Vertical asymptote)}, \ y = x - 3 \text{ (Slant asymptote)}$$

Setting $f(x) = 0$, we get:

$$x^2 - 3x + 2 = 0 \Rightarrow (x-2)(x-1) = 0 \Rightarrow x = 2, \ 1$$

The $x-$intercepts are 1 and 2.

The first derivative is: $f'(x) = 1 - \dfrac{2}{x^2}$ When $x = \pm\sqrt{2}$, $f'(x) = 0$, the function has critical points: $f(\sqrt{2}) = 2\sqrt{2} - 3$ $f(-\sqrt{2}) = -2\sqrt{2} - 3$.

When $x = 0$, $f'(x)$ does not exist, here the function has vertical asymptote.

The second derivative is: $f''(x) = \dfrac{4}{x^3}$, which is non-zero.

At $x = 0$, $f''(0)$ is not defined but it is not a point of inflection.

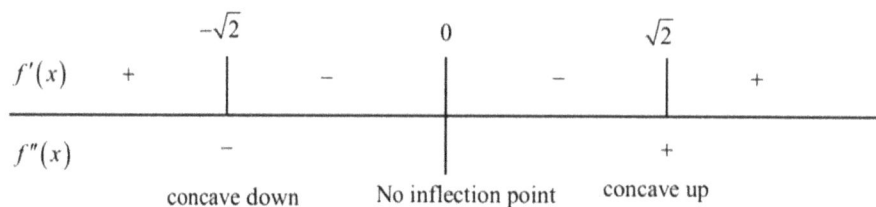

The graph of the function is given by:

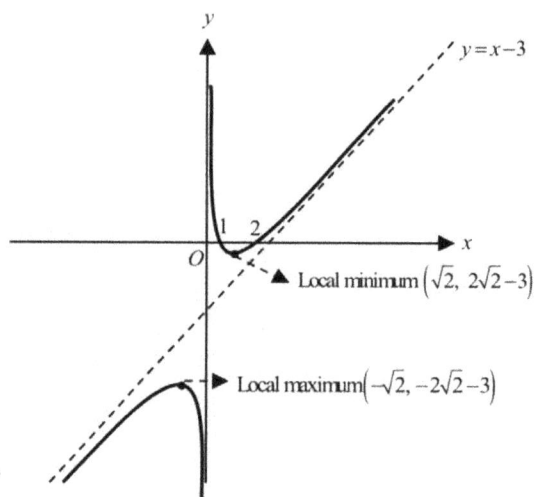

Example (5.7-1). Given the graph of f' in the figure below, find where the function f:

(a) has a horizontal tangent, (b) has its relative extrema, (c) is increasing or decreasing, (d) has a point of inflection, and (e) is concave upward or downward.

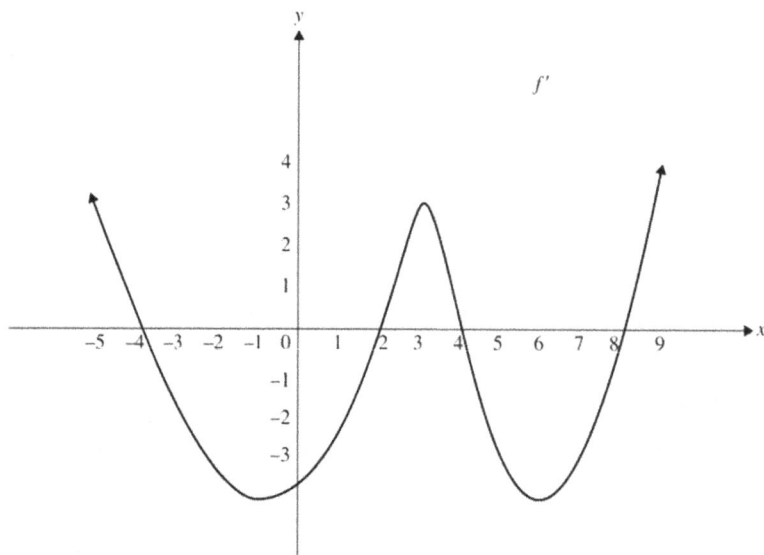

Solution.

(a) $f'(x) = 0$ at $x = -4$, 2, 4, 8. Thus f has horizontal tangents at these values.

(b) Summarize the information of f' on the following graph

$$f' \quad + \quad 0 \quad - \quad 0 \quad + \quad 0 \quad - \quad 0 \quad +$$

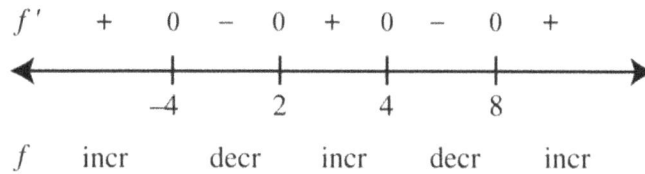

$$f \quad \text{incr} \quad \text{decr} \quad \text{incr} \quad \text{decr} \quad \text{incr}$$

The First Derivative Test indicates that f has relative maximums at $x = -4$ and 4 ; and f has relative minimums at $x = 2$ and 8.

(c) f is increasing on $(\infty, -4]$, $[2, 4]$, and $[8, \infty)$ and is decreasing on $[-4, -2]$ and $[4, 8]$.

(d) Summarize the information of f' on the following graph

$$f' \quad \text{decr} \quad \text{incr} \quad \text{decr} \quad \text{incr}$$

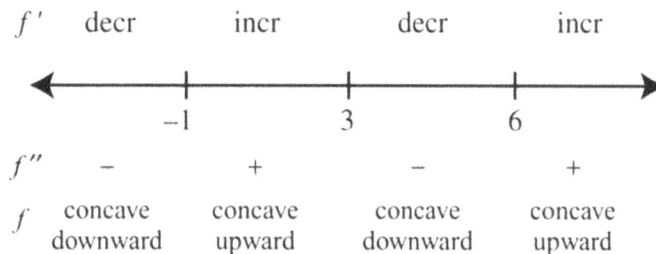

$$f'' \quad - \quad + \quad - \quad +$$

$$f \quad \begin{array}{c}\text{concave} \\ \text{downward}\end{array} \quad \begin{array}{c}\text{concave} \\ \text{upward}\end{array} \quad \begin{array}{c}\text{concave} \\ \text{downward}\end{array} \quad \begin{array}{c}\text{concave} \\ \text{upward}\end{array}$$

A change of concavity occurs at $x = -1$, 3, and 6. Since $f'(x)$ exists, f has a tangent at every point. Therefore, f has a point of inflection at $x = -1$, 3, and 6.

(e) The function f is concave upward on $(-1, 3)$ and $(6, \infty)$ and concave downward on $(-\infty, -1)$ and (3,6).

Example (5.7-2). A function f is continuous on the interval $[-4, 3]$ with $f(-4) = 6$ and $f(3) = 2$ and the following properties:

INTERVALS	(-4, -2)	x = -2	(-2, 1)	x = 1	(1, 3)
f'	–	0	–	undefined	+
f''	+	0	–	undefined	–

(a) Find the intervals on which f is increasing or decreasing.

(b) Find where f has its absolute extrema.

(c) Find where f has the points of inflection.

(d) Find the intervals where f is concave upward or downward.

(e) Sketch a possible graph of f.

Solution.

(a) The graph of f is increasing on $[1, 3]$ since $f'(x) > 0$ and decreasing on $[-4, -2]$ and $[-2, 1]$ since $f'(x) < 0$.

(b) At $x = -4$, $f(-4) = 6$. The function decreases until $x = 1$ and increases back to 2 at $x = 3$. Thus, f has its absolute maximum at $x = -4$ and its absolute minimum at $x = 1$.

(c) A change of concavity occurs at $x = -2$, and since $f''(-2) = 0$ and f'' changes its sign, thus f has a point of inflection at $x = -2$.

(d) The graph of f is concave upward on $(-4, -2)$ and concave downward on $(-2, 1)$ and $(1, 3)$.

(e) The graph of f is:

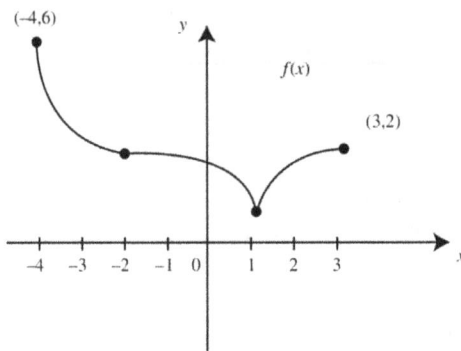

Example (5.7-3). Give the graph of $f''(x)$ in the figure below, determine the values of x at which the function f has a point of inflection.

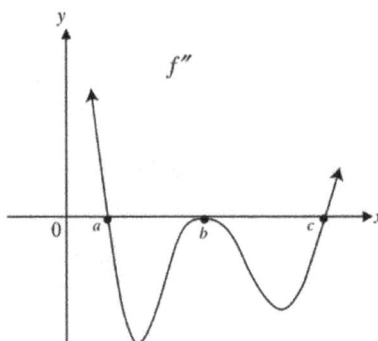

Solution. f has two points of inflection, one at $x = a$ and the other at $x = c$.

210

Example (5.7-4). The graph of f is shown in the figure below and f is twice differentiable. Which of the following has the largest value:

(A) $f(-1)$

(B) $f'(-1)$

(C) $f''(-1)$

(D) $f(-1)$ and $f'(-1)$

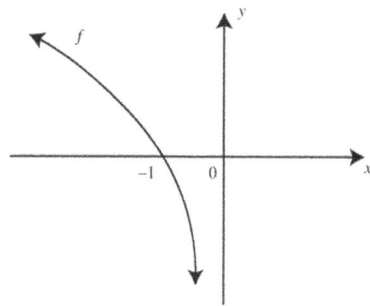

Solution.

$f(-1) = 0$; $f'(0) < 0$ since f is decreasing and $f''(-1) < 0$ since f is concave downward.

Thus, $f(-1)$ has the largest value, choice (A).

Example (5.8-1). Find the shortest distance between the point $A(19, 0)$ and the parabola $y = x^2 - 2x + 1$.

Solution.

Let $P(x, y)$ be the point on the parabola and let d represent the distance between points $P(x, y)$ and $A(19, 0)$.

Using the distance formula,

$$d = \sqrt{(x-19)^2 + (y-0)^2} = \sqrt{(x-19)^2 + (x^2 - 2x + 1)^2} = \sqrt{(x-19)^2 + (x-1)^4}$$

Let $L = d^2 = (x-19)^2 + (x-1)^4$. The domain of L is all real numbers.

The derivative of L with respect to x is:

$$\frac{dL}{dx} = 2(x-19) + 4(x-1)^3$$
$$= 2x - 38 + 4x^3 - 12x^2 + 12x - 4$$
$$= 4x^3 - 12x^2 + 14x - 42$$
$$= 2(2x^3 - 6x^2 + 7x - 21)$$

Let $\dfrac{dL}{dx} = 0$, we get:

$$2(2x^3 - 6x^2 + 7x - 21) = 2(x-3)(2x^2 + 7) = 0 \Rightarrow x = 3$$

Thus, the only critical point is at $x = 3$.

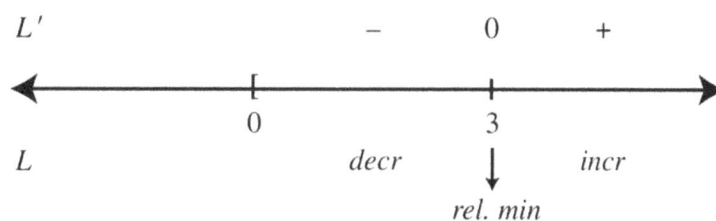

Since $x = 3$ is the only relative minimum point in the interval, it is the absolute minimum.

At $x = 3$, $d = \sqrt{(3 - 19)^2 + (3 - 1)^4} = \sqrt{16^2 + 16} = 4\sqrt{17}$

Example (5.8-2). A window is constructed by adjoining a semicircle and a rectangle. If the total perimeter is 24 feet, what is the radius of the semicircle that will maximize the area of the window?

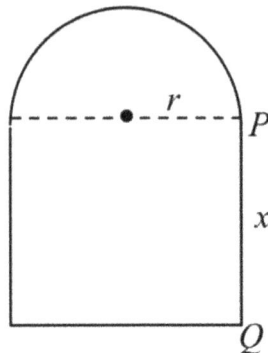

Solution.

If $PQ = x$, then the perimeter of the window is $2x + 2r + \pi r$.

The primary equation to be maximized: $A = 2rx + \dfrac{\pi r^2}{2}$

Constraints: $2x + 2r + \pi r = 24 \Rightarrow x = 12 - r - \dfrac{\pi}{2}r$, $r \in [0, 12]$

Then A can be expressed as:

$$A(r) = 2r(12 - r - \frac{\pi}{2}r) + \frac{\pi r^2}{2} = 24r - 2r^2 - \frac{\pi}{2}r^2$$

The derivative of $A(r)$ is given by:

$$A'(r) = 24 - 4r - \pi r = 0 \quad \Rightarrow \quad r = \frac{24}{\pi + 4}$$

Since $A''(r) = -4 - \pi < 0$, for all r, thus the window has maximum area at $r = \dfrac{24}{\pi + 4}$

Example (5.8-3). If an open box is to be made using a square sheet of tin, 20 inches by 20 inches, by cutting a square from each corner and folding the sides up, find the length of a side of the square being cut so that the box will have a maximum volume.

Solution.

Let x be the length of a side of the square to be cut from each corner.

The volume of the box is $V(x) = x(20 - 2x)^2 = 4x^3 - 80x^2 + 400x$, $\quad x \in [0, 10]$.

$V'(x) = 12x^2 - 160x + 400 = 4(x - 10)(3x - 10) = 0 \Rightarrow x = 10, \dfrac{10}{3}$.

Since $V(10) = 0$ and $V\left(\dfrac{10}{3}\right) = 592.59$, we need to test $V\left(\dfrac{10}{3}\right)$ only.

The second derivative is $V''(x) = 24x - 160 \Rightarrow V''\left(\dfrac{10}{3}\right) = -80 < 0$.

Thus, $V\left(\dfrac{10}{3}\right) = 592.59$ is the maximum volume, the length of a side of the square to be cut is

$$x = \frac{10}{3}.$$

11.6 Unit 6 Solutions

Example (6.1-1).

The graph shows the rate of change for the number of people in a museum t hours after it opens.

(a) How many people are in the museum after 5 hours?

(b) How many people are in the museum after 10 hours?

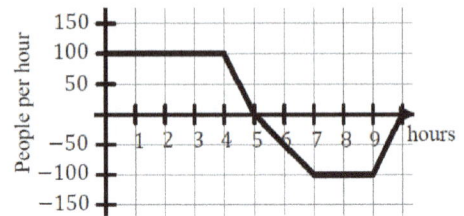

Solution.

(a) The number of people is actually the accumulated change from $t = 0$ to $t = 5$.

The Area $= \dfrac{1}{2} \times (4 + 5) \times 100 = 450$ people.

(b) The number of people is actually the accumulated change from $t = 0$ to $t = 10$.

In the first 5 hours, 450 people entered the museum. In the next 5 hours, the line is under the horizontal axis, which means that the number of people is decreasing.

The Signed Area $= 450 + \dfrac{1}{2}(2 + 5) \times (-100) = 450 - 350 = 100$ people.

Example (6.1-2).

The amount of water in a small pond is changing at the rate modeled in the graph, where the rate is measured in cubic inches per hour.

(a) How much water has been lost during the first 3 hours?

(b) How much water has been gained during the first 10 hours?

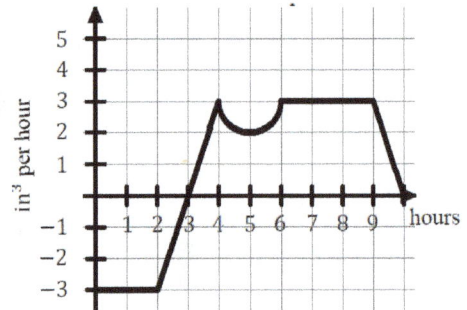

Solution.

(a) The water loss is equivalent to the area bounded by the curve from $t = 0$ to $t = 3$.

The Area $= \dfrac{1}{2} \times (2 + 3) \times (-3) = -7.5$ in^3.

Thus, 7.5in^3 water has been lost during the first 3 hours.

(b) The number of water gained is actually the accumulated change from $t = 0$ to $t = 10$.

The Signed Area $= -7.5 + \left(\dfrac{1}{2}(5 + 7) \times 3 - \dfrac{1}{2}\pi \cdot 1^2 \right) = \dfrac{21 - \pi}{2}$ in^3.

Example (6.2-1). Evaluate $\displaystyle\int x^5 - 6x^2 + x - 1 \ dx$.

Solution. The integral is given by:

$$\int x^5 - 6x^2 + x - 1 \, \mathrm{d}x = \frac{1}{6}x^5 - 2x^3 + \frac{1}{2}x^2 - x + C.$$

Example (6.2-2). Evaluate $\int 1 - \dfrac{1}{\sqrt[3]{x^4}} \, \mathrm{d}x$.

Solution. The integral is given by:

$$\int 1 - \frac{1}{\sqrt[3]{x^4}} \, \mathrm{d}x = \int 1 - x^{-\frac{4}{3}} \, \mathrm{d}x = x + 3x^{-\frac{1}{3}} + C = x + \frac{3}{\sqrt[3]{x}} + C.$$

Example (6.2-3). If $\dfrac{dy}{dx} = 3x^2 + 2$, and the point $(0, -1)$ lies on the graph of y, find y.

Solution. Since $\dfrac{dy}{dx} = 3x^2 + 2$, then y is an antiderivative of $\dfrac{dy}{dx}$.

$$\int 3x^2 + 2 \ dx = x^3 + 2x + C.$$

The point $(0, -1)$ is on the graph y.

Thus, $-1 = 0^3 + 2 \cdot 0 + C \ \Rightarrow \ C = -1$. Therefore, $y = x^3 + 2x - 1$.

Example (6.2-4). Evaluate $\displaystyle\int \dfrac{3x^3 + x^2 - 1}{x^2} \ dx$.

Solution. The integral is given by:

$$\int \frac{3x^3 + x^2 - 1}{x^2} \ dx = \int 3x + 1 - \frac{1}{x^2} \ dx = \int 3x + 1 - x^{-2} \ dx$$
$$= \frac{3}{2}x^2 + x + \frac{1}{x} + C.$$

Example (6.2-5). Evaluate $\displaystyle\int \sqrt{x}\left(x^2 - 3\right) \ dx$.

Solution. The integral is given by:

$$\int \sqrt{x}\left(x^2 - 3\right) \ dx = \int x^{\frac{1}{2}}\left(x^2 - 3\right) \ dx = \int x^{\frac{5}{2}} - 3x^{\frac{1}{2}} \ dx$$
$$= \frac{2}{7}x^{\frac{7}{2}} - 2x^{\frac{1}{3}} + C.$$

Example (6.2-6). Evaluate $\displaystyle\int \dfrac{3x^2 + x - 1}{x^2} \ dx$.

Solution. The integral is given by:

$$\int \frac{3x^2 + x - 1}{x^2} \ dx = \int 3 + \frac{1}{x} - \frac{1}{x^2} \ dx = \int 3 + \frac{1}{x} - x^{-2} \ dx$$
$$= 3x + \ln|x| + \frac{1}{x} + C.$$

Example (6.2-7). Evaluate $\displaystyle\int x - \csc x \cot x \ dx$.

Solution. The integral is given by:

$$\int x - \csc x \cot x \ \mathrm{d}x = \frac{1}{2}x^2 + \csc x + C.$$

Example (6.2-8). Evaluate $\displaystyle\int \sec x(\tan x - \sec x) \ \mathrm{d}x$.

Solution. The integral is given by:

$$\int \sec x(\tan x - \sec x) \ \mathrm{d}x = \int \sec x \tan x - \sec^2 x \ \mathrm{d}x$$
$$= \sec x + \tan x + C.$$

(Note: Sometimes we need to use the trigonometric identities to help integrate functions.)

Example (6.2-9). Evaluate $\int \tan^2 x \, dx$.

Solution. The integral is given by:

$$\int \tan^2 x \, dx = \int \sec^2 x - 1 \, dx = \tan x - x + C.$$

Example (6.2-10). Evaluate $\int \dfrac{\sin x}{1 - \sin^2 x} \, dx$.

Solution. The integral is given by:

$$\int \frac{\sin x}{1 - \sin^2 x} \, dx = \int \frac{\sin x}{\cos^2 x} \, dx = \int \frac{\sin x}{\cos x} \cdot \frac{\sin x}{\cos x} \, dx$$
$$= \int \sec x \tan x \, dx = \sec x + C.$$

Example (6.2-11). Evaluate $\int \dfrac{3}{2x + 1} \, dx$.

Solution. The integral is given by:

$$\int \frac{3}{2x + 1} \, dx = 3 \int \frac{1}{2x + 1} \, dx = \frac{3}{2} \ln |2x + 1| + C.$$

Example (6.2-12). Evaluate $\int \dfrac{3}{e^x} \, dx$.

Solution. The integral is given by:

$$\int \frac{3}{e^x} \, dx = \int 3e^{-x} \, dx = 3 \int e^{-x} \, dx = -3e^{-x} + C.$$

Example (6.2-13). Evaluate $\int \sqrt[3]{(4x - 5)^2} \, dx$.

Solution. The integral is given by:

$$\int \sqrt[3]{(4x - 5)^2} \, dx = \int (4x - 5)^{\frac{2}{3}} \, dx = \frac{3}{5} \cdot \frac{1}{4}(4x - 5)^{\frac{5}{3}} = \frac{3}{20}(4x - 5)^{\frac{5}{3}}.$$

Example (6.2-14). Evaluate $\int \sin^2 x \, dx$.

Solution. The integral is given by:

$$\int \sin^2 x \, dx = \int \frac{1 - \cos 2x}{2} \, dx = \int \frac{1}{2} - \frac{1}{2} \cos 2x \, dx = \frac{1}{2}x - \frac{1}{4} \sin 2x + C.$$

(Note: We used the *reduction formula* for $\sin^2 x$, the same method applies for integrating $\cos^2 x$.)

Example (6.2-15). Evaluate $\int \dfrac{1}{x^2 + 9} \, dx$.

Solution. The integral is given by:

$$\int \frac{1}{x^2 + 9} \, dx = \frac{1}{9} \int \frac{1}{\left(\frac{x}{3}\right)^2 + 1} \, dx = \frac{3}{9} \tan^{-1} x + C = \frac{1}{3} \tan^{-1}\left(\frac{x}{3}\right) + C$$

Example (6.3-1). Evaluate $\int \left(x^2 + 4\right)^2 (2x) \, dx$.

Solution. Let $u = x^2 + 4$, then $du = 2x \, dx$. Therefore:

$$\int \left(x^2 + 4\right)^2 (2x) \, dx = \int u^2 \, du = \frac{u^3}{3} + C = \frac{\left(x^2 + 4\right)^3}{3} + C$$

Example (6.3-2). Evaluate $\int \sqrt{2x + 1} \, dx$.

Solution. Let $u = 2x + 1$, then $du = 2 \, dx \;\Rightarrow\; dx = \dfrac{1}{2} \, du$. Therefore:

$$\int \sqrt{2x + 1} \, dx = \int u^{\frac{1}{2}} \cdot \frac{1}{2} \, du$$
$$= \frac{1}{2} \cdot \frac{2}{3} u^{\frac{3}{2}} = \frac{1}{3} u^{\frac{3}{2}} + C$$
$$= \frac{1}{3}(2x + 1)^{\frac{3}{2}} + C$$
$$= \frac{1}{3}\sqrt{(2x + 1)^3} + C.$$

Example (6.3-3). Evaluate $\int \dfrac{x^2}{(x^3 - 8)^5} \, dx$.

Solution. Let $u = x^3 - 8$, then $du = 3x^2 \, dx \;\Rightarrow\; \dfrac{1}{3} du = x^2 \, dx$. Therefore:

$$\int \frac{x^2}{(x^3 - 8)^5} \, dx = \int \frac{1}{u^5} \cdot \frac{1}{3} \, du = \frac{1}{3} \cdot \left(-\frac{1}{4}\right) u^{-4} = -\frac{1}{12}\left(x^3 - 8\right)^{-4} + C.$$

Example (6.3-4). Evaluate $\displaystyle\int \frac{4x + 6}{x^2 + 3x}\, dx$.

Solution. Let $u = x^2 + 3x$, then $du = (2x + 3))\, dx$. Therefore:

$$\int \frac{4x + 6}{x^2 + 3x}\, dx = \int \frac{2(2x + 3)}{x^2 + 3x}\, dx = \int \frac{2}{u}\, du = 2\ln|u| + C = 2\ln\left|x^2 + 3x\right| + C.$$

Example (6.3-5). Evaluate $\displaystyle\int \tan x\, dx$.

Solution. $\displaystyle\int \tan x\, dx = \int \frac{\sin x}{\cos x}\, dx$. Let $u = \cos x$, then $du = -\sin x\, dx$. Therefore:

$$\int \frac{\sin x}{\cos x}\, dx = -\int \frac{-\sin x}{\cos x}\, dx = -\int \frac{1}{u}\, du$$
$$= -\ln|u| + C = -\ln|\cos x| + C = \ln|\sec x| + C.$$

Example (6.3-6). By considering $\sec x = \dfrac{\sec x(\sec x + \tan x)}{\sec x + \tan x}$, evaluate $\displaystyle\int \tan x\, dx$.

Solution.

Since $\displaystyle\int \tan x\, dx = \int \frac{\sec x(\sec x + \tan x)}{\sec x + \tan x}\, dx = \int \frac{\sec^2 x + \sec x \tan x}{\sec x + \tan x}\, dx$.

Let $u = \sec x + \tan x$, then $du = (\sec x \tan x + \sec^2 x)\, dx$. Therefore:

$$\int \frac{\sec^2 x + \sec x \tan x}{\sec x + \tan x}\, dx = \int \frac{1}{u}\, du = \ln|u| + C = \ln|\sec x + \tan x| + C.$$

Example (6.3-7). Evaluate $\displaystyle\int x\sqrt{2x + 1}\, dx$.

Solution. Let $u = 2x + 1$, then $x = \dfrac{1}{2}(u - 1) \Rightarrow dx = \dfrac{1}{2}du$. Therefore:

$$\int x\sqrt{2x + 1}\, dx = \frac{1}{2}(u - 1) \cdot u^{\frac{1}{2}} \cdot \frac{1}{2}du$$
$$= \frac{1}{4}\int u^{\frac{2}{3}} - u^{\frac{1}{2}}\, du$$
$$= \frac{1}{4}\left(\frac{2}{5}u^{\frac{5}{2}} - \frac{2}{3}u^{\frac{3}{2}}\right) + C$$
$$= \frac{1}{10}(2x + 1)^{\frac{5}{2}} - \frac{1}{6}(2x + 1)^{\frac{3}{2}} + C.$$

Example (6.3-8). Evaluate $\displaystyle\int 2x^2 \cos\left(x^3\right)\,\mathrm{d}x$.

Solution. Let $u = x^3$, then $\mathrm{d}u = 3x^2\,\mathrm{d}x \Rightarrow \dfrac{1}{3}\mathrm{d}u = x^2\mathrm{d}x$. Therefore:

$$
\begin{aligned}
\int 2x^2 \cos\left(x^3\right)\,\mathrm{d}x &= 2\int \cos u \cdot \frac{1}{3}\,\mathrm{d}u \\
&= \frac{2}{3}\int \cos u\,\mathrm{d}u \\
&= \frac{2}{3}\sin u + C = \frac{2}{3}\sin\left(x^3\right) + C.
\end{aligned}
$$

Example (6.3-9). Evaluate $\displaystyle\int \sin^2 3x \cdot \cos 3x\,\mathrm{d}x$.

Solution. Let $u = \sin 3x$, then $\mathrm{d}u = 3\cos 3x\,\mathrm{d}x \Rightarrow \dfrac{1}{3}\mathrm{d}u = \cos 3x\mathrm{d}x$. Therefore:

$$
\begin{aligned}
\int \sin^2 3x \cdot \cos 3x\,\mathrm{d}x &= 2\int u^2 \cdot \frac{1}{3}\,\mathrm{d}u \\
&= \frac{1}{3}\int u^2\,\mathrm{d}u \\
&= \frac{1}{9}u^3 + C = \frac{1}{9}\sin^3 3x + C.
\end{aligned}
$$

Example (6.3-10). Evaluate $\displaystyle\int xe^{3x^2}\,\mathrm{d}x$.

Solution. Let $u = 3x^2$, then $\mathrm{d}u = 6x\,\mathrm{d}x \Rightarrow \dfrac{1}{6}\mathrm{d}u = x\mathrm{d}x$. Therefore:

$$
\begin{aligned}
\int xe^{3x^2}\,\mathrm{d}x &= \int e^{3x^2} \cdot x\,\mathrm{d}x \\
&= \frac{1}{6}\int e^u\,\mathrm{d}u \\
&= \frac{1}{6}e^u + C = \frac{1}{6}e^{3x^2} + C.
\end{aligned}
$$

Example (6.3-11). Evaluate $\displaystyle\int \frac{1}{x\ln(x^2)}\,\mathrm{d}x$.

Solution. Since $\displaystyle\int \frac{1}{x\ln(x^2)}\,\mathrm{d}x = \int \frac{1}{2x\ln x}\,\mathrm{d}x$

Let $u = \ln x$, then $du = \dfrac{1}{x}\, dx$. Therefore:

$$\int \frac{1}{x \ln(x^2)}\, dx = \int \frac{1}{2x \ln x}\, dx$$

$$= \frac{1}{2} \int \frac{1}{x \ln x}\, dx = \frac{1}{2} \int \frac{1}{u}\, du = \frac{1}{2} \ln|u| + C = \frac{1}{2} \ln|\ln x| + C.$$

Example (6.3-12). Evaluate $\displaystyle \int \frac{\ln x}{3x}\, dx$.

Solution. Let $u = \ln x$, then $du = \dfrac{1}{x}\, dx$. Therefore:

$$\int \frac{\ln x}{3x}\, dx = \int \frac{u}{3}\, du = \frac{1}{6}u^2 + C = \frac{1}{6}\left(\ln x\right) + C.$$

Example (6.3-13). Evaluate $\displaystyle \int \frac{1}{\sqrt{9 - 4x^2}}\, dx$.

Solution. The integrand can be re-written as:

$$\int \frac{1}{\sqrt{9 - 4x^2}}\, dx = \int \frac{1}{3\sqrt{1 - \frac{4}{9}x^2}}\, dx = \frac{1}{3} \int \frac{1}{\sqrt{1 - \frac{4}{9}x^2}}\, dx$$

Let $u = \dfrac{2}{3}x$, then $du = \dfrac{2}{3}\, dx \Rightarrow dx = \dfrac{3}{2}\, du$. Therefore:

$$\frac{1}{3} \int \frac{1}{\sqrt{1 - \frac{4}{9}x^2}}\, dx = \frac{1}{3} \int \frac{1}{\sqrt{1 - u^2}} \cdot \frac{3}{2}\, du$$

$$= \frac{1}{2} \int \frac{1}{\sqrt{1 - u^2}}\, du = \frac{1}{2} \sin^{-1}(u) + C = \frac{1}{2} \sin^{-1}\left(\frac{2}{3}x\right) + C.$$

Example (6.3-14). Evaluate $\displaystyle \int \frac{x(x - 2)}{(x - 1)^3}\, dx$.

Solution. Let $u = x - 1$, then $x = u + 1$ and $du = dx$. Therefore:

$$\int \frac{x(x - 2)}{(x - 1)^3}\, dx = \int \frac{(u + 1)(u - 1)}{u^3}\, du = \int \frac{u^2 - 1}{u^3}\, du$$

$$= \int \frac{1}{u} - u^{-3}\, du = \ln|u| - \frac{1}{-2}u^{-2} + C$$

$$= \ln|u| + \frac{1}{2u^2} + C$$

$$= \ln|x - 1| + \frac{1}{2(x - 1)^2} + C.$$

Example (6.4-1). Find the approximate area under the curve of $f(x) = \sqrt{x}$ from $x = 4$ to $x = 9$ using 5 right-endpoint rectangles, and determine whether it is an overestimate or underestimate.

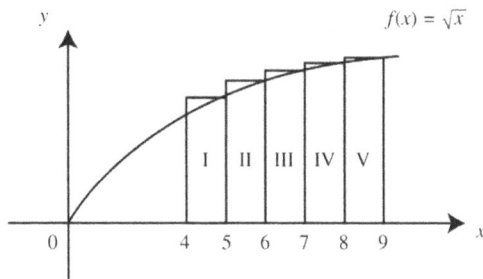

Solution.

Let Δx be the width of each rectangle. Then $\Delta x = \dfrac{9 - 4}{5} = 1$ and $x_k = 4 + k\Delta x = 4 + k$.

Then the total area using *right-endpoint* is given by:

$$\sum_{k=1}^{5} f(x_k)\Delta x = f(x_1)\Delta x + f(x_2)\Delta x + f(x_3)\Delta x + f(x_4)\Delta x + f(x_5)\Delta x$$

$$= f(5) \cdot 1 + f(6) \cdot 1 + f(7) \cdot 1 + f(8) \cdot 1 + f(9) \cdot 1$$

$$= \sqrt{5} + \sqrt{6} + \sqrt{7} + \sqrt{8} + 3 = 13.160$$

According to the graph, this is an overestimate.

Example (6.4-2). The function f is positive and continuous on $[1, 9]$. Selected values of f are given:

x	1	2	3	4	5	6	7	8	9
$f(x)$	1	1.41	1.73	2	2.37	2.45	2.65	2.83	3

Using 4 midpoint rectangles, approximate the area under the curve of f for $x = 1$ to $x = 9$.

Solution.

Let Δx be the width of each rectangle. Then $\Delta x = \dfrac{9 - 1}{4} = 2$ and $x_k = 4 + k\Delta x = 4 + k$.

Then the total area using *midpoint* is given by:

$$\sum_{k=1}^{4} f\left(\frac{x_{k-1} + x_k}{2}\right)\Delta x = f(\frac{1+3}{2})\Delta x + f(\frac{3+5}{2})\Delta x + f(\frac{5+7}{2})\Delta x + f(\frac{7+9}{2})\Delta x$$

$$= f(2) \cdot 2 + f(4) \cdot 2 + f(6) \cdot 2 + f(8) \cdot 2$$

$$= 2 \times (1.41 + 2 + 2.45 + 2.83) = 17.38$$

Thus the area under the curve is approximately 17.38.

Example (6.4-3). Find the approximate area under the curve of from to , using 4 trapezoids.

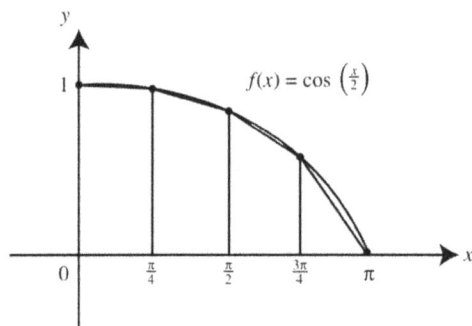

Solution. Since $n = 4$, $\Delta x = \dfrac{\pi - 0}{4} = \dfrac{\pi}{4}$

Then the total area using *trapezodial approximation* is given by:

$$\text{Area} \approx \frac{1}{2} \cdot \frac{\pi}{4} \left[\cos(0) + 2\cos\left(\frac{\pi/4}{2}\right) + 2\cos\left(\frac{\pi/2}{2}\right) + 2\cos\left(\frac{3\pi/4}{2}\right) + 2\cos\left(\frac{\pi}{2}\right) \right]$$

$$= \frac{\pi}{8} \left[\cos(0) + 2\cos\left(\frac{\pi}{8}\right) + 2\cos\left(\frac{\pi}{4}\right) + 2\cos\left(\frac{3\pi}{8}\right) + 2\cos\left(\frac{\pi}{2}\right) \right]$$

$$= \frac{\pi}{8} \left[1 + 2(0.9239) + 2 \cdot \frac{\sqrt{2}}{2} + 2(0.3827) + 0 \right] = 1.9743$$

Thus the area under the curve is approximately 1.9743.

Example (6.5-1). Write the summation notation for the integral $\displaystyle\int_1^7 \sqrt{x}\ \mathrm{d}x$ and $\displaystyle\int_0^6 \sqrt{1+x}\ \mathrm{d}x$

Solution.

The summation of $\displaystyle\int_1^7 \sqrt{x}\ \mathrm{d}x$ is given by:

$$\lim_{n\to\infty} \sum_{k=1}^{n} \sqrt{1 + \frac{6k}{n}} \left(\frac{6}{n}\right)$$

The summation of $\displaystyle\int_0^6 \sqrt{1+x}\ \mathrm{d}x$ is given by:

$$\lim_{n\to\infty} \sum_{k=1}^{n} \sqrt{0 + 1 + \frac{6k}{n}} \left(\frac{6}{n}\right) = \lim_{n\to\infty} \sum_{k=1}^{n} \sqrt{1 + \frac{6k}{n}} \left(\frac{6}{n}\right)$$

(Note: The integral notation for a summation is not unique, when the interval is not restricted)

Example (6.5-2). Write the integral notation for $\displaystyle\lim_{n\to\infty} \sum_{k=1}^{n} \cos\left(4 + \frac{6k}{n}\right)\left(\frac{6}{n}\right)$, which lower limit $=$
4

Solution.

Since $\displaystyle\lim_{n\to\infty} \sum_{k=1}^{n} f\left(a + \frac{b-a}{n}k\right) \cdot \left(\frac{b-a}{n}\right) = \int_a^b f(x)\, dx$, then we have $a = 4$, $b = 10$

Then we have $f\left(a + \dfrac{b-a}{n}k\right) = \cos\left(4 + \dfrac{10-4}{n}k\right)$, thus $f(x) = \cos x$

Therefore, the integral is: $\displaystyle\int_4^{10} \cos x\, dx$

Example (6.5-3). Consider the expression

$$\lim_{n\to\infty} \left(\frac{2}{n}\right)\left(\frac{1}{\dfrac{2}{n}+3} + \frac{1}{\dfrac{4}{n}+3} + \frac{1}{\dfrac{6}{n}+3} + \cdots + \frac{1}{\dfrac{2n}{n}+3}\right)$$

Assuming the lower limit a is 0, write a definite integral that represents the above expression.

Solution.

Since $\dfrac{b-a}{n} = \dfrac{2}{n}$ and $a = 0$, then $b = 2$.

Then we have $f\left(a + \dfrac{b-a}{n}k\right) = \dfrac{1}{3 + \dfrac{2-0}{n}k} = \dfrac{1}{3 + \left(0 + \dfrac{2-0}{n}k\right)}$, thus $f(x) = \dfrac{1}{3+x}$

Therefore, the integral is: $\displaystyle\int_0^2 \frac{1}{3+x}\, dx$

Example (6.5-4). Which of the following integrals are equal to $\displaystyle\lim_{n\to\infty} \sum_{k=1}^{n} \left(-1 + \frac{4k}{n}\right)^2 \frac{4}{n}$

I. $\displaystyle\int_{-1}^{3} x^2\, dx$ II. $\displaystyle\int_0^4 (-1+x)^2\, dx$ III. $\displaystyle\int_0^1 4(-1+4x)^2\, dx$

Solution. By rewriting all of them to summation notation:

$$\int_{-1}^{3} x^2\, dx = \lim_{n\to\infty} \sum_{k=1}^{n} \left(-1 + \frac{4k}{n}\right)^2 \frac{4}{n}$$

$$\int_0^4 (-1+x)^2\, dx = \lim_{n\to\infty} \sum_{k=1}^{n} \left(-1 + \left(0 + \frac{4k}{n}\right)\right)^2 \frac{4}{n} = \lim_{n\to\infty} \sum_{k=1}^{n} \left(-1 + \frac{4k}{n}\right)^2 \frac{4}{n}$$

$$\int_0^1 4(-1+4x)^2\, dx = \lim_{n\to\infty} \sum_{k=1}^{n} \left[4 \cdot \left(-1 + 4\left(0 + \frac{k}{n}\right)\right)\right]^2 \frac{1}{n} = \lim_{n\to\infty} \sum_{k=1}^{n} \left(-1 + \frac{4k}{n}\right)^2 \frac{4}{n}$$

Thus, all of them are equal to the summation.

Example (6.6-1). Evaluate $\displaystyle\int_1^4 \frac{x^3 - 8}{\sqrt{x}}\, dx$.

Solution. The definite integral is given by:

$$\int_1^4 \frac{x^3 - 8}{\sqrt{x}}\, dx = \int_1^4 \left(x^{\frac{5}{2}} - 8x^{-\frac{1}{2}} \right) dx = \frac{2}{7}x^{\frac{7}{2}} - 16x^{\frac{1}{2}} \Big|_1^4$$

$$= \left(\frac{2}{7}(4)^{\frac{7}{2}} - 16(4)^{\frac{1}{2}} \right) - \left(\frac{2}{7}(1)^{\frac{7}{2}} - 16(1)^{\frac{1}{2}} \right) = \frac{142}{7}$$

Example (6.6-2). Evaluate $\displaystyle\int_{\frac{\pi}{4}}^{\frac{\pi}{2}} \csc^2 3t\, dt$.

Solution. The definite integral is given by:

$$\int_{\frac{\pi}{4}}^{\frac{\pi}{2}} \csc^2 3t\, dt = -\frac{1}{3}\cot 3t \Big|_1^4$$

$$= -\frac{1}{3}\left[\cot \frac{3\pi}{2} - \cot \frac{3\pi}{4} \right] = -\frac{1}{3}[0 - (-1)] = -\frac{1}{3}$$

The integration can also be solved by u-substitution, but we need to pay attention to the boundaries.

Let $u = 3t \Rightarrow du = 3\, dt$ or $dt = \frac{1}{3}\, du$, then the integral becomes:

$$\int_{t=\frac{\pi}{4}}^{t=\frac{\pi}{2}} \csc^2 3t\, dt = \int_{u=\frac{3}{4}\pi}^{u=\frac{3}{2}\pi} \frac{1}{3}\csc^2 u\, du = \frac{1}{3}\int_{\frac{3}{4}\pi}^{\frac{3}{2}\pi} \csc^2 u\, du$$

$$= \frac{1}{3}\cdot\left(-\cot u \big|_{\frac{3}{2}\pi}^{\frac{3}{4}\pi} \right) = \frac{1}{3}(0 - 1) = -\frac{1}{3}$$

Example (6.6-3). Evaluate $\displaystyle\int_0^{\sqrt{\ln 2}} xe^{-x^2}\, dx$.

Solution. Let $u = -x^2$, then $du = -2x\, dx \Rightarrow -\frac{1}{2}\, du = x\, dx$

$$\int_0^{\sqrt{\ln 2}} xe^{-x^2}\, dx = \int_0^{-\ln 2} -\frac{1}{2}e^u\, du = \frac{1}{2}\int_{-\ln 2}^0 e^u\, du$$

$$= \frac{1}{2}\left(e^u \big|_{-\ln 2}^0 \right) = \frac{1}{2}\left(e^0 - e^{-\ln 2} \right) = \frac{1}{2}\left(1 - \frac{1}{2} \right) = \frac{1}{4}$$

Example (6.6-4). Evaluate $\displaystyle\int_1^{\sqrt[4]{3}} \frac{x}{1 + x^4}\, dx$.

Solution. Let $u = x^2$, then $\mathrm{d}u = 2x\,\mathrm{d}x \;\Rightarrow\; \dfrac{1}{2}\,\mathrm{d}u = x\,\mathrm{d}x$

$$\int_1^{\sqrt[4]{3}} \frac{x}{1+x^4}\,\mathrm{d}x = \int_1^{\sqrt{3}} \frac{1}{2}\cdot\frac{1}{1+u^2}\,\mathrm{d}u = \frac{1}{2}\int_1^{\sqrt{3}} \frac{1}{1+u^2}\,\mathrm{d}u$$

$$= \frac{1}{2}\left(\tan^{-1} u\Big|_0^{\sqrt{3}}\right) = \frac{1}{2}\left(\tan^{-1}\sqrt{3}\right) - \tan^{-1}(0)\right) = \frac{1}{2}\left(\frac{\pi}{3} - 0\right) = \frac{\pi}{6}$$

Example (6.6-5). Evaluate $\displaystyle\int_1^4 |3x - 6|\,\mathrm{d}x$.

Solution. The integrad can be piewise defined as: $f(x) = |3x - 6| = \begin{cases} 3x - 6, & x \geq 2 \\ 6 - 3x, & x < 2 \end{cases}$

Then the definite integral is given by:

$$\int_1^4 |3x - 6|\,\mathrm{d}x = \int_1^2 (6 - 3x)\,\mathrm{d}x + \int_2^4 (3x - 6)\,\mathrm{d}x$$

$$= \left(6x - \frac{3}{2}x^2\right)\Big|_1^2 + \left(\frac{3}{2}x^2 - 6x\right)\Big|_2^4$$

$$= \left[(12 - 6) - \left(6 - \frac{3}{2}\right)\right] + [(24 - 24) - (6 - 12)] = \frac{15}{2}$$

Example (6.6-6). If $h(x) = \displaystyle\int_3^x \sqrt{t+1}\,\mathrm{d}t$ find $h'(8)$

Solution. By the FTC Part 1, we have:

$$h'(x) = \sqrt{x+1} \;\Rightarrow\; h'(8) = \sqrt{8+1} = 3$$

Example (6.6-7). Find $\dfrac{\mathrm{d}}{\mathrm{d}x}\displaystyle\int_0^{\cos^2(2x)} \sqrt{t}\,\mathrm{d}t$

Solution. Let $u = \cos^2(2x)$. By the FTC Part 1, we have:

$$\frac{\mathrm{d}}{\mathrm{d}x}\int_0^{\cos^2(2x)} \sqrt{t}\,\mathrm{d}t = \frac{\mathrm{d}}{\mathrm{d}x}\int_0^{u(x)} \sqrt{t}\,\mathrm{d}t = \sqrt{u}\cdot u'$$

$$= \sqrt{\cos^2(2x)}\cdot 2\cos(2x)(-2\sin 2x)$$

$$= -4\sin(2x)\cos(2x)\,|\cos(2x)|$$

Example (6.6-8). Find $\dfrac{\mathrm{d}y}{\mathrm{d}x}$ if $y = \displaystyle\int_{x^2}^1 \sin t\,\mathrm{d}t$

Solution. Let $u = x^2$. By the FTC Part 1, we have:

$$\frac{dy}{dx} = -\frac{d}{dx} \int_1^{x^2} \sin t \; dt = -\frac{d}{dx} \int_1^{u(x)} \sin t \; dt = -\sin u \cdot u'$$
$$= -\sin\left(x^2\right) \cdot 2x = -2x \sin\left(x^2\right)$$

Example (6.7-1). Evaluate $\int xe^{-x} \; dx$.

Solution. Let $u = x$ and $dv = e^{-x} \; dx$, then $du = dx$ and $v = -e^{-x}$. Then:

$$\int xe^{-x} \; dx = uv - \int v \; du$$
$$-xe^{-x} - \int -e^{-x} \; dx = -xe^{-x} - e^{-x} + C = -e^{-x}(x+1) + C.$$

Example (6.7-2). Evaluate $\int x^2 \ln x \; dx$.

Solution. Let $u = \ln x$ and $dv = x^2 \; dx$, then $du = \frac{1}{x} \; dx$ and $v = \frac{x^3}{3}$. Then:

$$\int x^2 \ln x \; dx = uv - \int v \; du$$
$$= \frac{x^3 \ln x}{3} - \int \frac{x^2}{3} \; dx = \frac{x^3 \ln x}{3} - \frac{x^3}{9} + C.$$

Example (6.7-3). Evaluate $\int \ln x \; dx$.

Solution. Let $u = \ln x$ and $dv = 1 \cdot dx$, then $du = \frac{1}{x} \; dx$ and $v = x$. Then:

$$\int x^2 \ln x \; dx = uv - \int v \; du$$
$$= x \ln x - \int 1 \; dx = x \ln x - x + C.$$

Example (6.7-4). Evaluate $\int e^x \cos x \; dx$.

Solution. Let $u = e^x$ and $dv = \cos x \; dx$, then $du = e^x \; dx$ and $v = \sin x$. Then:

$$I = \int e^x \cos x \; dx = uv - \int v \; du$$
$$= e^x \sin x - \int e^x \sin x \; dx$$

Let $u = e^x$ and $dv = \sin x \, dx$, then $du = e^x \, dx$ and $v = -\cos x$. Then:

$$I = e^x \sin x - \int e^x \sin x \, dx = e^x \sin x - \left(-e^x \cos x - \int -e^x \cos x \, dx \right)$$

$$I = e^x \sin x + e^x \cos x - I$$

Moving the I to the left hand side, we get:

$$2I = e^x \sin x + e^x \cos x \;\Rightarrow\; I = \frac{1}{2} e^x (\sin x + \cos x) + C.$$

Example (6.7-5). By considering $\sec^3 x = \sec x \cdot \sec^2 x$, evaluate $\int \sec^3 x \, dx$.

Solution.

For functions with *same type*, we may choose the one having *harder* integration as u.

Let $u = \sec x$ and $dv = \sec^2 x \, dx$, then $du = \sec x \tan x \, dx$ and $v = \tan x$. Then:

$$I = \int \sec x \cdot \sec^2 x \, dx = uv - \int v \, du$$

$$= \sec x \tan x - \int \sec x \tan^2 x \, dx$$

Here we may use the identity $\tan^2 x = \sec^2 x - 1$, the the integration becomes:

$$I = \sec x \tan x - \int \sec x (\sec^2 x - 1) \, dx$$

$$= \sec x \tan x - \int \sec^3 x - \sec x \, dx$$

$$= \sec x \tan x - \int \sec^3 x \, dx + \int \sec x \, dx$$

$$= \sec x \tan x - I + \int \sec x \, dx$$

Moving the I to the left hand side, we get:

$$2I = \sec x \tan x + \int \sec x \, dx$$

$$I = \frac{1}{2} \left(\sec x \tan x + \int \sec x \, dx \right)$$

$$= \frac{1}{2} \left(\sec x \tan x + \ln |\sec x + \tan x| \right) + C.$$

Example (6.7-6). Evaluate $\int x^2 \sin x \, dx$.

Solution.

Method 1: Let $u = x$, $dv = \sin x\ dx$, then $du = 2x\ dx$, $v = -\cos x$

$$\int x^2 \sin x\ dx = uv - \int v\ du$$

$$= -x^2 \cos x - \int -2x \cos x\ dx = -x^2 \cos x + \int 2x \cos x\ dx$$

Let $u = 2x$, $dv = \cos x\ dx$, then $du = 2\ dx$, $v = \sin x$

$$\int x^2 \sin x\ dx = -x^2 \cos x + \int 2x \cos x\ dx$$

$$= -x^2 \cos x + 2x \sin x - \int 2 \sin x\ dx$$

$$= -x^2 \cos x + 2x \sin x + 2 \cos x + C.$$

Method 2: By the tabular method

$+$	x^2	$\sin x$	
$-$	$2x$	$-\cos x$	$-x^2 \cos x$
$+$	2	$-\sin x$	$-2x \sin x$
$-$	0	$\cos x$	$2 \cos x$

Thus $\int x^2 \sin x\ dx = -x^2 \cos x + 2x \sin x + 2 \cos x + C.$

Example (6.8-1). Evaluate $\displaystyle\int \frac{1}{x^2 + 3x - 4}\ dx$.

Solution. By the partial fractions decomposition, we have:

$$\int \frac{1}{x^2 + 3x - 4}\ dx = \int \frac{1}{(x-1)(x+4)}\ dx = \int \frac{A}{x-1} + \frac{B}{x+4}\ dx$$

Equating the numerator for both sides: $A(x+4) + B(x-1) \equiv 1$

Let $x = 1$, then $5A = 1 \Rightarrow A = \dfrac{1}{5}$

Let $x = -4$, then $-5B = 1 \Rightarrow B = -\dfrac{1}{5}$

Thus, the integration is given by:

$$\int \frac{1}{5(x-1)} - \frac{1}{5(x+4)}\ dx = \frac{1}{5} \ln|x-1| - \frac{1}{5} \ln|x+4| + C.$$

Example (6.8-2). Evaluate $\displaystyle\int \frac{5x^2 + 7x}{(x^2-1)(2x+1)}\ dx$.

Solution. By the partial fractions decomposition, we have:

$$\int \frac{5x^2 + 7x}{(x^2-1)(2x+1)}\ dx = \int \frac{x(5x+7)}{(x+1)(x-1)(2x+1)}\ dx = \int \frac{A}{x+1} + \frac{B}{x-1} + \frac{D}{2x+1}\ dx$$

Equating the numerator for both sides:

$$A(x-1)(2x+1) + B(x+1)(2x+1) + D(x+1)(x-1) \equiv x(5x+7)$$

Let $x = -1$, then $2A = -2 \Rightarrow A = -1$

Let $x = 1$, then $6B = 12 \Rightarrow B = 2$

Let $x = -\dfrac{1}{2}$, then $-\dfrac{3}{4}C = \dfrac{9}{4} \Rightarrow B = 3$

Thus, the integration is given by:

$$\int -\frac{1}{x+1} + \frac{2}{x-1} + \frac{3}{2x+1} \; \mathrm{d}x = -\ln|x+1| + 2\ln|x-1| + \frac{3}{2}\ln|2x+1| + C.$$

Example (6.8-3). Evaluate $\displaystyle\int \frac{x+1}{(x-1)^2} \; \mathrm{d}x$.

Solution. By the partial fractions decomposition, we have:

$$\int \frac{x+1}{(x-1)^2} \; \mathrm{d}x = \int \frac{A}{x-1} + \frac{B}{(x-1)^2} \; \mathrm{d}x$$

Equating the numerator for both sides: $A(x-1) + B \equiv x+1$

Let $x = 1$, then $B = 2$

Let $x = 2$, then $A + B = 3 \Rightarrow A = 1$ (you can let x be any other values)

Thus, the integration is given by:

$$\int \frac{1}{x-1} + \frac{2}{(x-1)^2} \; \mathrm{d}x = \ln|x-1| - \frac{2}{x-1} + C.$$

Example (6.8-4). Evaluate $\displaystyle\int \frac{x+1}{(x-1)(x-2)^2} \; \mathrm{d}x$.

Solution. By the partial fractions decomposition, we have:

$$\int \frac{x+1}{(x-1)(x-2)^2} \; \mathrm{d}x = \int \frac{A}{x-1} + \frac{B}{x-2} + \frac{D}{(x-2)^2} \; \mathrm{d}x$$

Equating the numerator for both sides:

$$A(x-2)^2 + B(x-1)(x-2) + C(x-1) \equiv x+1$$

Let $x = 1$, then $A = 2$

Let $x = 2$, then $C = 3$

Let $x = 0$, then $4A + 2B - C = 1 \Rightarrow B = -2$ (you can let x be any other values)

Thus, the integration is given by:

$$\int \frac{x+1}{(x-1)(x-2)^2} \; \mathrm{d}x = \int \frac{2}{x-1} - \frac{2}{x-2} + \frac{3}{(x-2)^2} \; \mathrm{d}x$$
$$= 2\ln|x-1| - 2\ln|x-2| - \frac{3}{x-2} + C.$$

Example (6.8-5). Evaluate $\displaystyle\int \frac{2x^2 + x + 1}{(x-1)(x^2+1)} \; \mathrm{d}x$.

Solution. By the partial fractions decomposition, we have:

$$\int \frac{2x^2 + x + 1}{(x-1)(x^2+1)}\, dx = \int \frac{A}{x-1} + \frac{Bx+D}{x^2+1}\, dx$$

Equating the numerator for both sides: $A(x^2+1) + (Bx+D)(x-1) \equiv 2x^2 + x + 1$

Let $x = 1$, then $2A = 4 \Rightarrow A = 2$

Let $x = 0$, then $A - D = 1 \Rightarrow D = 1$

Let $x = -1$, then $2A - 2(-B+D) = 2 \Rightarrow 4 - 2(-B+1) = 2 \Rightarrow B = 0$

Thus, the integration is given by:

$$\int \frac{2x^2 + x + 1}{(x-1)(x^2+1)}\, dx = \int \frac{2}{x-1} + \frac{1}{x^2+1}\, dx$$
$$= 2\ln|x-1| + \tan^{-1} x + C.$$

Example (6.8-6). Evaluate $\displaystyle\int \frac{3-x}{(x+1)(x^2+3)}\, dx$.

Solution. By the partial fractions decomposition, we have:

$$\int \frac{3-x}{(x+1)(x^2+3)}\, dx = \int \frac{A}{x+1} + \frac{Bx+D}{x^2+3}\, dx$$

Equating the numerator for both sides: $A(x^2+3) + (Bx+D)(x+1) \equiv 3 - x$

Let $x = -1$, then $4A = 4 \Rightarrow A = 1$

Let $x = 0$, then $3A + D = 3 \Rightarrow D = 0$

Let $x = 1$, then $4A + 2B + 2D = 2 \Rightarrow 4 + 2B = 2 \Rightarrow B = -1$

Thus, the integration is given by:

$$\int \frac{3-x}{(x+1)(x^2+3)}\, dx = \int \frac{1}{x+1} - \frac{x}{x^2+3}\, dx$$
$$= \ln|x+1| + \frac{1}{2}\ln(x^2+3) + C.$$

Example (6.8-7). Evaluate $\displaystyle\int \frac{2x^2 + 5x - 11}{x^2 + 2x - 3}\, dx$.

Solution. The integrand can be decomposed as:

$$\int \frac{2x^2 + 5x - 11}{x^2 + 2x - 3}\, dx = \int 2 + \frac{x-5}{(x+3)(x-1)}\, dx = \int 2 + \frac{A}{x+3} + \frac{B}{x-1}\, dx$$

Equating the numerator for both sides: $A(x-1) + B(x+3) \equiv x - 5$

Let $x = -3$, then $-4A = -8 \Rightarrow A = 2$ Let $x = 1$, then $4B = -4 \Rightarrow B = -1$

Thus, the integration is given by:

$$\int \frac{2x^2 + 5x - 11}{x^2 + 2x - 3}\, \mathrm{d}x = \int 2 + \frac{2}{x+3} - \frac{1}{x-1}\, \mathrm{d}x$$
$$= 2x + 2\ln|x+3| - \ln|x-1| + C.$$

Example (6.8-8). Evaluate $\displaystyle\int \frac{1}{x^2 + 2x + 5}\, \mathrm{d}x$.

Solution. The integrand can be decomposed as:

$$\int \frac{1}{x^2 + 2x + 5}\, \mathrm{d}x = \int \frac{1}{(x+1)^2 + 2^2}\, \mathrm{d}x = \frac{1}{4}\int \frac{1}{\left(\frac{x+1}{2}\right)^2 + 1}\, \mathrm{d}x$$

Let $u = \dfrac{x+1}{2} = \dfrac{1}{2}x + \dfrac{1}{2}$, then $\mathrm{d}u = \dfrac{1}{2}\mathrm{d}x \;\Rightarrow\; \mathrm{d}x = 2\, \mathrm{d}u$

$$\frac{1}{4}\int \frac{1}{\left(\frac{x+1}{2}\right)+1}\, \mathrm{d}x = \frac{1}{4}\int \frac{2}{u^2+1}\, \mathrm{d}x = \frac{1}{2}\tan^{-1} u + C = \frac{1}{2}\tan^{-1}\left(\frac{x+1}{2}\right) + C.$$

Example (6.9-1). Evaluate $\displaystyle\int_1^\infty \frac{1}{x}\, \mathrm{d}x$.

Solution. The integration is given by:

$$\int_1^\infty \frac{1}{x}\, \mathrm{d}x = \lim_{b\to\infty} \int_1^b \frac{1}{x}\, \mathrm{d}x$$
$$= \lim_{b\to\infty} \ln x\big|_1^b$$
$$= \lim_{b\to\infty} (\ln b - 0) = \infty$$

Thus, the integral diverges.

Example (6.9-2). Evaluate $\displaystyle\int_0^\infty \frac{1}{x^2+1}\, \mathrm{d}x$.

Solution. The integration is given by:

$$\int_0^\infty \frac{1}{x^2+1}\, \mathrm{d}x = \lim_{b\to\infty} \int_0^b \frac{1}{x^2+1}\, \mathrm{d}x$$
$$= \lim_{b\to\infty} \tan^{-1} x\big|_0^b$$
$$= \lim_{b\to\infty} (\tan^{-1} b - 0) = \frac{\pi}{2}$$

Thus, the integral converges to $\dfrac{\pi}{2}$.

Example (6.9-3). Evaluate $\displaystyle\int_{-\infty}^\infty xe^{-x^2}\, \mathrm{d}x$.

Solution. The integration is given by:

$$\int_{-\infty}^{\infty} xe^{-x^2}\,dx = \int_{-\infty}^{0} xe^{-x^2}\,dx + \int_{0}^{\infty} xe^{-x^2}\,dx$$

$$= \lim_{a \to -\infty} \int_{a}^{0} xe^{-x^2}\,dx + \lim_{b \to \infty} \int_{0}^{b} xe^{-x^2}\,dx$$

$$= \lim_{a \to -\infty} \left(-\frac{1}{2}e^{=x^2}\Big|_{a}^{0} \right) + \lim_{b \to \infty} \left(-\frac{1}{2}e^{=x^2}\Big|_{b}^{0} \right)$$

$$= \lim_{a \to -\infty} \left(-\frac{1}{2} + \frac{1}{2}e^{-a^2} \right) + \lim_{b \to \infty} \left(-\frac{1}{2}e^{-b^2+\frac{1}{2}} \right) = \frac{1}{2} + \frac{1}{2} = 0$$

Thus, the integral converges to 0.

Example (6.9-4). Evaluate $\int_{0}^{\frac{\pi}{2}} \frac{\cos x}{\sqrt{1 - \sin x}}\,dx$.

Solution. Since $f(x)$ has an infinite discontinuity at $x = \dfrac{\pi}{2}$, the improper integral is:

$$\int_{0}^{\frac{\pi}{2}} \frac{\cos x}{\sqrt{1 - \sin x}}\,dx = \lim_{k \to \frac{\pi}{2}^{-}} \int_{0}^{k} \frac{\cos x}{\sqrt{1 - \sin x}}\,dx$$

$$= \lim_{k \to \frac{\pi}{2}^{-}} \left(-2\sqrt{1 - \sin x}\Big|_{0}^{k} \right)$$

$$= \lim_{k \to \frac{\pi}{2}^{-}} \left(-2\sqrt{1 - \sin k} + 2 \right) = 2$$

Thus, the integral converges to 2.

11.7   Unit 7 Solutions

Example (7.1-1). Find the general solution of the differentiation equation $\dfrac{\mathrm{d}v}{\mathrm{d}t} = 2t - e^{-t}$.

Solution. By definite integral:

$$v = \int 2t - e^{-t} \ \mathrm{d}t = t^2 + e^{-t} + C$$

Thus, the general solution is: $y = t^2 + e^{-t} + C$

Example (7.1-2). Find a solution of the differentiation equation $\dfrac{\mathrm{d}y}{\mathrm{d}x} = x \sin\left(x^2\right)$, with $y(0) = -1$.

Solution. By definite integral: Let $u = x^2$, then $\mathrm{d}u = 2x \ \mathrm{d}x$, and $\dfrac{1}{2} \ \mathrm{d}u = x \ \mathrm{d}x$

$$y = \int x \sin\left(x^2\right) \ \mathrm{d}x = \frac{1}{2} \int \sin u \ \mathrm{d}u$$
$$= -\frac{1}{2} \cos u + C = -\frac{1}{2} \cos\left(x^2\right) + C.$$

Substitute with the given condition $y(0) = -1$:

$$-1 = -\frac{1}{2} \cos(0) + C \Rightarrow -1 = -\frac{1}{2} + C \Rightarrow C = -\frac{1}{2}$$

Thus, the particular solution is $y = -\dfrac{1}{2} \cos\left(x^2\right) - \dfrac{1}{2}$.

Example (7.1-3). Find the general solution of the differentiation equation $\dfrac{\mathrm{d}^2 y}{\mathrm{d}x^2} = \dfrac{1}{x}$ with $x > 0$.

Solution. By definite integral:
$$\frac{\mathrm{d}y}{\mathrm{d}x} = \int \frac{1}{x} \ \mathrm{d}x = \ln x + C$$

The we integrate again to get the solution:

$$y = \int \ln x \ \mathrm{d}x = x \ln x - x + Cx + C_2$$
$$= x \ln x + C_1 x + C_2, \quad (\text{where } C_1 = C - 1)$$

Thus, the general solution is: $y = x \ln x + C_1 x + C_2$

Example (7.1-4). If $\dfrac{\mathrm{d}^2 y}{\mathrm{d}x^2} = 2x + 1$ and at $x = 0$, $y' = -1$ and $y = 3$, find a solution of the differential equation.

Solution. By definite integral:

$$y' = \frac{\mathrm{d}y}{\mathrm{d}x} = \int 2x + 1 \ \mathrm{d}x = x^2 + x + C_1$$

At $x = 0$, $y' = -1$, then $0 + C_1 = C_1 = -1$, and $y' = x^2 + x - 1$

The we integrate again to get the solution:

$$y = \int x^2 + x - 1 \ dx = \frac{1}{3}x^3 + \frac{1}{2}x^2 - x + C_2$$

At $x = 0$, $y = 3$, then $0 + C_2 = C_2 = 3$

Thus, the general solution is: $y\frac{1}{3}x^3 + \frac{1}{2}x^2 - x + 3$

Example (7.2-1). Sketch the slope field of the differential equation $\dfrac{dy}{dx} = x^2(y-2)$ on the coordinate plane with vertices points with $-1 \le x \le 1$ and $0 \le y \le 5$.

Solution. At each point $(0, y)$ and $(x, 2)$: we have zero slope.

	$x = -1$	$x = 0$	$x = 1$
$y = 0$	-2	0	-2
$y = 1$	-1	0	-1
$y = 2$	0	0	0
$y = 3$	1	0	1
$y = 4$	2	0	2
$y = 5$	3	0	3

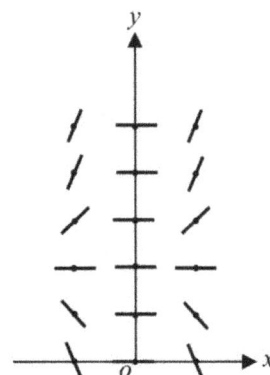

Example (7.2-2). The figure below shows a slope field for one of the differential equations given below. Which of the following equation does the figure indicate.

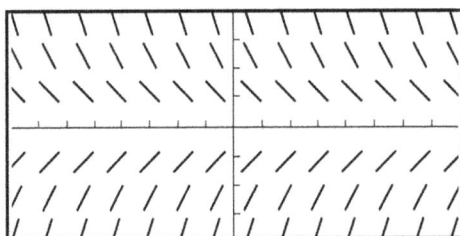

(A) $\dfrac{dy}{dx} = 2x$ (B) $\dfrac{dy}{dx} = -2x$ (C) $\dfrac{dy}{dx} = y$ (D) $\dfrac{dy}{dx} = -y$ (E) $\dfrac{dy}{dx} = x + y$

Solution. The correct choice is (D) $\dfrac{dy}{dx} = -y$, since the slope depends on y only.

Example (7.2-3). Which of the following could be a slope field for the differential equation $\dfrac{dy}{dx} = x^2 + y^2$?

(A)

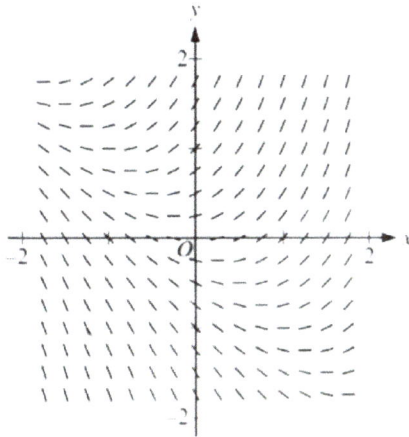

(B)

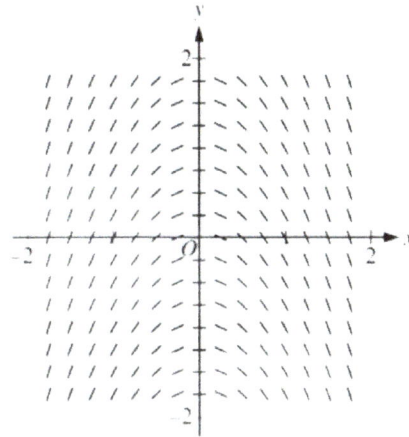

(C)

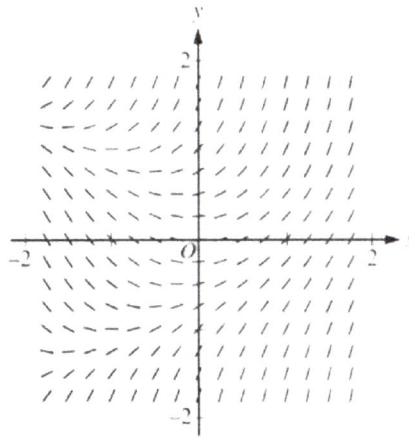

(D)

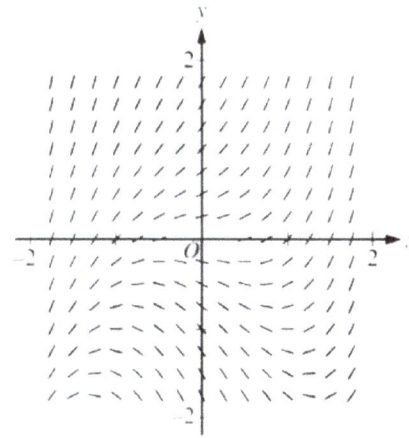

Solution. The correct choice is (D).

Example (7.2-4). Match each slope field in the figures below with the proper differential equation from the following set. The particular solution that goes through $(0,0)$ has been sketched in.

(1) $\dfrac{dy}{dx} = \cos x$ (2) $\dfrac{dy}{dx} = 2x$ (3) $\dfrac{dy}{dx} = 3x^2 - 3$ (4) $\dfrac{dy}{dx} = -\dfrac{\pi}{2}$

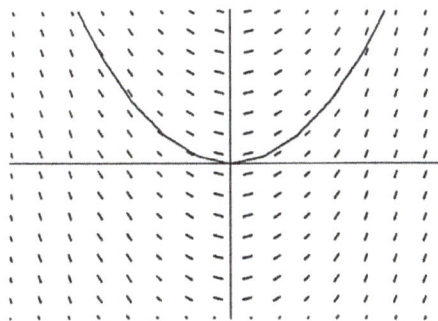

$[-2,2] \times [-2,2]$

(A)

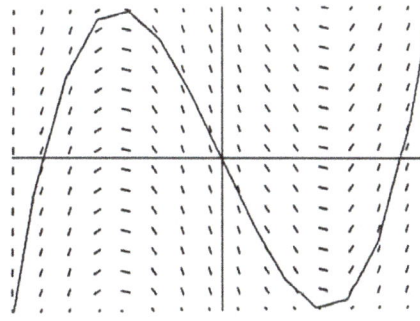

$[-2,2] \times [-2,2]$

(B)

237

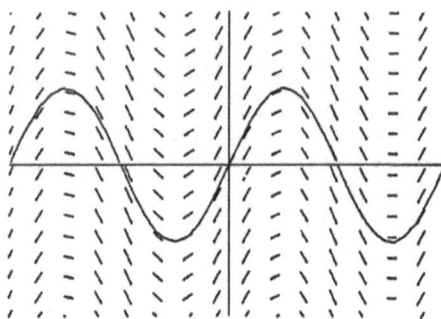

$[-2\pi, 2\pi] \times [-2,2]$

(C)

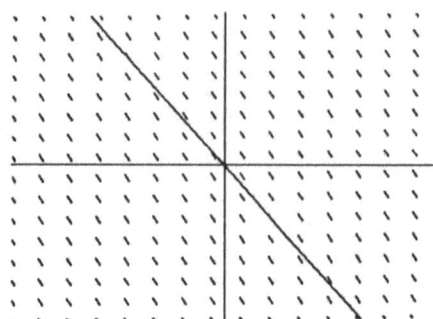

$[-2,2] \times [-2,2]$

(D)

Solution.

(1) is shown in figure (C); (2) is shown in figure (A)

(3) is shown in figure (B); (4) is shown in figure (D)

Example (7.3-1). Use Euler's method with a step size of $h = 1/4$ (or $\Delta x = 1/4$) to approximate $y(2)$ if $\dfrac{dy}{dx} = y + 1$ and point $(1,1)$ belongs to the graph of the solution of the differential equation.

Solution. By $y_{n+1} = y_n + \Delta x \cdot y_n{}'$, we can make the following table:

Points	$\Delta x = h$ (step size)	$\dfrac{dy}{dx} = y+1$	$\Delta y = \Delta x \cdot \left(\dfrac{dy}{dx}\right)$	$y + \Delta y$
$(1,1)$	0.25	2	$0.25(2) = 0.5$	1.5
$(1.25, 1.5)$	0.25	2.5	0.625	2.125
$(1.5, 2.125)$	0.25	3.125	0.78125	2.90625
$(1.75, 2.90625)$	0.25	3.90625	0.9765625	3.8828125
$(2, 3.8828125)$				

Therefore, $y(2) \approx 3.8828125$

238

Example (7.3-2). Use Euler's method with step size $h = 0.2$ to approximate $y(1.6)$ if $\dfrac{dy}{dx} = \dfrac{x+y}{x}$ and $y(1) = 2$.

Solution. By $y_{n+1} = y_n + \Delta x \cdot y_n'$, we can make the following table:

Points	$\Delta x = h$ (step size)	$\dfrac{dy}{dx} = \dfrac{x+y}{x}$	$\Delta y = \Delta x \cdot \left(\dfrac{dy}{dx}\right)$	$y + \Delta y$
$(1, 2)$	0.2	3	0.6	2.6
$(1.2, 2.6)$	0.2	3.17	0.634	3.234
$(1.4, 3.234)$	0.2	3.31	0.662	3.896
$(1.6, 3.896)$				

Therefore, $y(1.6) \approx 3.896$

Example (7.4-1). Given $\dfrac{dy}{dx} = 4x^3 y^2$ and $y(1) = -\dfrac{1}{2}$, solve the differential equation.

Solution.

1. Separate the variables: $\dfrac{1}{y^2}\, dy = 4x^3\, dx$

2. Integrate both sides: $\displaystyle\int \dfrac{1}{y^2}\, dy = \int 4x^3\, dx$

3. Find Antiderivatives: $-\dfrac{1}{y} = x^4 + C$

4. Get General Solution: $y = -\dfrac{1}{x^4 + C}$

5. Substitute $x = 1$, $y = -\dfrac{1}{2}$ $\Rightarrow$ $-\dfrac{1}{2} = -\dfrac{1}{1+C}$ $\Rightarrow$ $C = 1$;

Thus, the particular solution to this equation is: $y = -\dfrac{1}{x^4 + 1}$

Note: You may also substitute $x = 1$, $y = -\dfrac{1}{2}$ after you find the antiderivatives of both sides, then

$$2 = 1^4 + C \ \Rightarrow\ C = 1 \ \Rightarrow\ -\dfrac{1}{y} = x^4 + 1 \ \Rightarrow\ y = -\dfrac{1}{x^4 + 1}$$

Example (7.4-2). Find the general solution of the differential equation $\dfrac{dy}{dx} = \dfrac{2xy}{x^2+1}$.

Solution.

1. Separate the variables: $\dfrac{1}{y}\,dy = \dfrac{2x}{x^2+1}\,dx$

2. Integrate both sides: $\displaystyle\int \dfrac{1}{y}\,dy = \int \dfrac{2x}{x^2+1}\,dx$ (let $u = x^2+1 \;\Rightarrow\; du = 2x\,dx$)

3. Find Antiderivatives: $\ln|y| = \ln\left(x^2+1\right) + C_1$

4. Get General Solution by raised to the e's power:

$$e^{\ln|y|} = e^{\ln\left(x^2+1\right)+C_1}$$
$$|y| = e^{C_1} \cdot \left(x^2+1\right)$$
$$y = \pm e^{C_1} \cdot \left(x^2+1\right)$$

Since $\pm e^{C_1}$ can be considered as an arbitrary constant, we can substitute it with C.

Therefore, the general solution is $y = C\left(x^2+1\right)$

Example (7.4-3). Find the particular solution of the differential equation $\dfrac{dy}{dx} = -2y^2$, with initial condition $y(0) = -\dfrac{1}{2}$.

Solution. Though there are no x on the right hand side, yet we can still separate the variables.

1. Separate the variables: $-\dfrac{1}{y^2}\,dy = 2\,dx$

2. Integrate both sides: $\displaystyle\int -\dfrac{1}{y^2}\,dy = \int 2\,dx$

3. Find Antiderivatives: $\dfrac{1}{y} = 2x + C$

4. Get General Solution: $y = \dfrac{1}{2x+C}$

5. Substitute $x = 0$, $y = -\dfrac{1}{2} \;\Rightarrow\; -\dfrac{1}{2} = \dfrac{1}{0+C} \;\Rightarrow\; C = -2$;

Thus, the particular solution to this equation is: $y = \dfrac{1}{2x-2}$.

Example (7.5-1). Show that the exponential model can be expressed as the differential equation $\dfrac{dy}{dt} = ky$, where $t \geq 0$, $y \geq 0$, and find the particular solution with initial condition $y(0) = y_0$

Solution. Since the differential equation is separable, so we can solve by the following:

1. Separate the variables: $\dfrac{1}{y}\,dy = k\,dt$

2. Integrate both sides: $\displaystyle\int \dfrac{1}{y}\,dy = \int k\,dt$

3. Find Antiderivatives: $\ln y = kt + c_1$

4. Get General Solution: $y = e^{kt+c_1} = e^{c_1} \cdot e^{kt} = Ce^{kt}$

Therefore, the general solution is $y = Ce^{kt}$.

By substituting $t = 0$, $y = y_0$, then $y_0 = Ce^0 = C$

Thus, the particular solution can be given by $y = y_0 \cdot e^{kt}$: If $k > 0$, then y is increasing when t increases. This is an exponential growth.

If $k < 0$, then y is decreasing when t increases. This is an exponential decay.

Example (7.5-2). The radioactive isotope Indium-111 is often used for diagnosis and imaging in nuclear medicine. Its half life is 2.8 days. What was the initial mass of the isotope before decay, if the mass in 2 weeks was 5 g.

Solution. The mass of isotope follows the exponential decay model by the function:

$$y = y_0 \cdot e^{-kt}$$

The half life is given by: $kt = \ln 2 \;\Rightarrow\; 2.8k = \ln 2 \;\Rightarrow\; k = \dfrac{\ln 2}{2.8}$

Substituting $t = 14$ into the modeling equation, we get:

$$5 = y_0 \cdot e^{-\frac{\ln 2}{2.8} \cdot 14} = y_0 \cdot e^{-5\ln 2} = \frac{y_0}{32} \;\Rightarrow\; y_0 = 160.$$

Example (7.5-3). The rate of growth of population of flies is proportional to the size of population. In an experiment, it was observed that there were 200 flies after the second day and 1000 flies after the fourth day. How many flies were there in the original population?

Solution. The population follows the exponential growth model by the function:

$$y = y_0 \cdot e^{kt}$$

At $t = 2$, $200 = y_0 \cdot e^{2k}$;   At $t = 4$, $1000 = y_0 \cdot e^{4k}$ Divide the two equations, we get:

$$\frac{1000}{200} = \frac{y_0 e^{4k}}{y_0 e^{2k}} = e^{2k} \;\Rightarrow\; e^{2k} = 5 \;\Rightarrow\; k = \frac{\ln 5}{2}, \; e^{2k} = 5$$

Substituting $e^{2k} = 5$ into the first equation, we get:

$$200 = y_0 \cdot 5 \;\Rightarrow\; y_0 = 40$$

We may also use the formulas in the previous COROLLARY:

$$k = \frac{\ln y_2 - \ln y_1}{t_2 - t_1} = \frac{\ln(1000) - \ln(200)}{4 - 2} = \frac{\ln 5}{2}$$
$$\ln y_0 = \frac{t_2 \ln y_1 - t_1 \ln y_2}{t_2 - t_1} = \frac{4\ln(200) - 2\ln(1000)}{4 - 2} = \frac{\ln(1600)}{2} = \ln(40)$$
$$\Rightarrow y_0 = 40$$

Example (7.6-1). If the population of a kind of cell is modelled by $\dfrac{dP}{dt} = 20P(400 - P)$, find the population when the population grows the fastest.

Solution. When $P = 200$, the population grows the fastest.

Example (7.6-2). The growing of a population is modeled by the following function: (P increases according to the logistic model):

$$\frac{dP}{dt} = \frac{4}{5}P(1 - \frac{P}{20})$$

(1) Find $P(t)$ if $P(0) = 5$.

(2) What is $\lim_{t \to \infty} P(t)$

(3) For what value of P is the population growing the fastest?

(4) For what value of t is the population growing the fastest?

Solution.

(1) From the equation: $k = \frac{4}{5}$ and $M = 20$.

　　The solution is $P = \frac{M}{1 + Ae^{-kt}} = \frac{20}{1 + Ae^{-0.8t}}$ and $P(0) = \frac{20}{1 + A}$

　　From $P(0) = 5$, $\frac{20}{1 + A} = 5 \Rightarrow A = 3$

　　Thus, $P(t) = \frac{20}{1 + 3e^{-0.8t}}$

(2) $\lim_{t \to \infty} P(t) = 20$, carrying capacity.

(3) At $P = 10$, the population is growing the fastest.

(4) At $t = t_0 = \frac{\ln A}{k} = \frac{\ln 3}{0.8} \approx 1.37$, the population is growing the fastest.

Example (7.6-3). A lake is stocked with 500 fish. If the population increases according to the logistic curve $y = \frac{10000}{1 + Ae^{-t/5}}$, where y is the fish population and t is measured in months.

(1) At what rate is the fish population changing at the end of the fifth month?

(2) After how many months is the population increasing the most rapidly?

Solution.

(1) In the logistic model, $M = 10000$ and $k = \frac{1}{5}$, then: $\frac{dy}{dt} = \frac{1}{5}y\left(1 - \frac{y}{10000}\right)$ When $t = 5$,

　　$y = \frac{10000}{1 + 19e^{-1}} \approx 1251.6$, then by the differential equation,

$$\frac{dy}{dt} = \frac{1}{5} \times 1251.6 \left(1 - \frac{1251.6}{10000}\right) \approx 219$$

(2) We may use the initial value $y(0)$ to determine A.

$$y(0) = \frac{10000}{1 + A} = 500 \Rightarrow A = 19$$

　　Since $A = 19$ and $k = \frac{1}{5}$, then $t = \frac{\ln A}{k} = \frac{\ln 19}{0.2} \approx 14.72$ months.

243

11.8   Unit 8 Solutions

Example (8.1-1). The graph of a function f is shown in figure. Find the average value of f on $[0, 4]$.

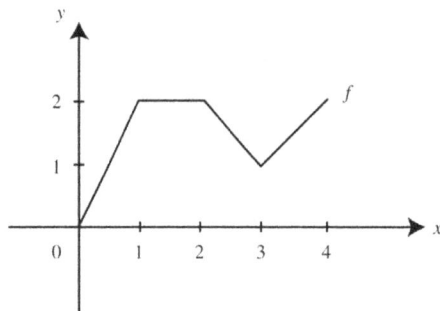

Solution. The average value is given by:

$$f_{\text{average}} = \frac{1}{4-0} \int_a^b f(x)\, dx = \frac{1}{4}\left(1 + 2 + \frac{3}{2} + \frac{3}{2}\right) = \frac{3}{2}$$

Example (8.1-2). Let f be the function that is defined for all real number x. f has following properties.

$$\text{(i) } f''(x) = 24x - 18 \qquad \text{(ii) } f'(1) = -6 \qquad \text{(iii) } f(2) = 0$$

(a) Find each x such that the line tangent to the graph of f at $(x, f(x)$ is horizontal.

(b) Find $f(x)$.

(c) Find the average value of f on the interval $1 \leq x \leq 3$.

Solution.

(a) $f'(x) = \displaystyle\int (24x - 18)\, dx = 12x^2 - 18x + C$, since $f'(1) = -6$, then

$f'(1) = 12 - 18 + C = -6 \Rightarrow C = 0$. Therefore $f'(x) = 12x^2 - 18x$.

Horizontal tangent occurs at $f'(x) = 0 \Rightarrow 12x^2 - 18x = 0$

$6x(2x - 3) = 0 \Rightarrow x = 0, \dfrac{3}{2}$

(b) $f(x) = \displaystyle\int 12x^2 - 18x\, dx = 4x^3 - 9x^2 + C$, since $f(2) = 0$, then

$f(2) = 32 - 36 + C = 0 \Rightarrow C = 4$

Therefore, $f(x) = 4x^3 - 9x^2 + 4$.

(c) The average value is given by:

$$f_{\text{avg}} = \frac{1}{3-1} \int_1^3 4x^3 - 9x^2 + 4\, dx = \frac{1}{2}\left(x^4 - 3x^3 + 4x\right)\Big|_1^3$$

$$= \frac{1}{2}[(81 - 81 + 12) - (1 - 3 + 4)] = \frac{1}{2}(12 - 2) = 5$$

Example (8.1-3). Let $f(x) = 10\pi x^2$ and $g(x) = k^2 \sin\left(\dfrac{\pi x}{2k}\right)$ for $k > 0$.

(a) What is the average value of f on $[1, 4]$?

(b) For what value of x will the average value of g on $[0, k]$ be equal to the average value of f on $[0, 4]$?

Solution.

(a) $f_{\text{avg}} = \dfrac{1}{4-1} \displaystyle\int_1^4 10\pi x^2 \, dx = \dfrac{10\pi}{3} \left(\dfrac{x^3}{3}\right)\bigg|_1^3 = \dfrac{10\pi}{3}\left(\dfrac{64}{3} - \dfrac{1}{3}\right) = 70\pi$

(b) The average of g on $[0, k]$ is

$$
g_{\text{avg}} = \dfrac{1}{k-0}\int_0^k k^2 \sin\left(\dfrac{\pi x}{2k}\right) dx = k\left(\dfrac{-\cos\frac{\pi x}{2k}}{\frac{\pi}{2k}}\right)\bigg|_0^k = -\dfrac{2k^2}{\pi}\left[\left(\dfrac{\pi x}{2k}\right)\right]\bigg|_0^k
$$

$$
= -\dfrac{2k^2}{\pi}\left[\cos\left(\dfrac{\pi}{2}\right) - \cos 0\right] = -\dfrac{2k^2}{\pi}(0-1) = \dfrac{2k^2}{\pi}
$$

Since $\dfrac{2k^2}{\pi} = 70\pi$, then $k^2 = 35\pi^2 \;\Rightarrow\; k = \sqrt{35}\pi$

Example (8.2-1). The graph of the velocity function of a moving particle is shown in the figure below. What is the total distance traveled by the particle during $0 \le t \le 12$?

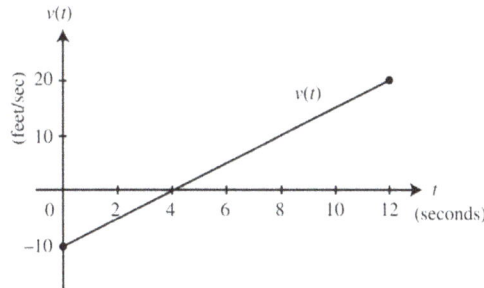

Solution. The total distance travelled is given by:

$$
\int_0^{12} |v(t)| \, dt = \int_0^4 -v(t) \, dt + \int_4^{12} v(t) \, dt \quad = \dfrac{1}{2}(4)(10) + \dfrac{1}{2}(8)(20) = 20 + 80 = 100 \text{ feet.}
$$

Example (8.2-2). The velocity of a moving particle on a line is $v(t) = t^2 + 3t - 10$ for $0 \le t \le 6$. Find (a) the displacement during $0 \le t \le 6$, and (b) the total distance traveled during $0 \le t \le 6$.

Solution.

(a) Displacement $= \displaystyle\int_0^6 v(t) \, dt = \int_0^6 t^2 + 3t - 10 \, dt = \dfrac{1}{3}t^3 + \dfrac{3}{2}t^2 - 10t \bigg|_0^6 = 66$

(b) Let $t^2 + 3t + 10 = 0 \Rightarrow t = -5$ or $t = 2 \Rightarrow |v(t)| = \begin{cases} -t^2 - 3t + 10, & 0 \le t < 2 \\ t^2 + 3t - 10, & t \ge 2 \end{cases}$

$$
\text{Total distance} = \int_0^6 |t^2 + 3t - 10| \, dt = \int_0^2 -t^2 - 3t + 10 \, dt + \int_2^6 t^2 + 3t - 10 \, dt
$$

$$
= \left(-\dfrac{1}{3}t^3 - \dfrac{3}{2}t^2 + 10t\right)\bigg|_0^2 + \left(\dfrac{1}{3}t^3 + \dfrac{3}{2}t^2 - 10t\right)\bigg|_2^{10} = \dfrac{266}{3}
$$

Example (8.2-3). The velocity function of a moving particle on a line is $v(t) = 3\cos(2t)$ for $0 \le t \le 2\pi$.

  (a) Determine when the particle is moving to the right.

  (b) Determine when the particle stops.

  (c) The total distance traveled by the particle during $0 \le t \le 2\pi$.

You might use your calculator to solve this problem.

Solution.

  (a) The particle is moving to the right when

      Let $v(t) = 0$, then $t = \dfrac{\pi}{4}, \dfrac{3\pi}{4}, \dfrac{5\pi}{4}$, and $\dfrac{7\pi}{4}$

      The particle is moving to the right when: $0 < t < \dfrac{3\pi}{4}, \dfrac{3\pi}{4} < t < \dfrac{5\pi}{4}, \dfrac{7\pi}{4} < t < 2\pi$.

  (b) The particle stops when $v(t) = 0$

      Thus, $t = \dfrac{\pi}{4}, \dfrac{3\pi}{4}, \dfrac{5\pi}{4}$, and $\dfrac{7\pi}{4}$

  (c) The total distance is given by:

$$\text{Total distance} = \int_0^{2\pi} |3\cos(2t)| \, dt = 12$$

Example (8.3-1). Find the area of the region bounded by the graph of $f(x) = x^2 - 1$, the lines $x = -2$ and $x = 2$, and the x-axis.

Solution. The Area is given by:

$$A = \int_{-2}^{-1} x^2 - 1 \, dx + \left| \int_{-1}^{1} x^2 - 1 \, dx \right| + \int_{1}^{2} x^2 - 1 \, dx$$

$$= \int_{-2}^{-1} x^2 - 1 \, dx - \int_{-1}^{1} x^2 - 1 \, dx + \int_{1}^{2} x^2 - 1 \, dx$$

$$= \left(\frac{1}{3}x^3 - 1 \right)\Big|_{-2}^{-1} - \left(\frac{1}{3}x^3 - 1 \right)\Big|_{-1}^{1} + \left(\frac{1}{3}x^3 - 1 \right)\Big|_{1}^{2}$$

$$= \left[\frac{2}{3} - \left(-\frac{2}{3} \right) \right] - \left[-\frac{2}{3} - \left(\frac{2}{3} \right) \right] + \left[\frac{2}{3} - \left(-\frac{2}{3} \right) \right]$$

$$= \frac{4}{3} - \left(\frac{4}{3} \right) + \frac{4}{3} = 4$$

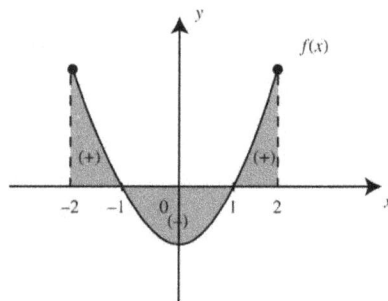

Example (8.3-2). Find the area of the region bounded by $x = y^2$, $y = -1$, $y = 3$ and the y-axis.

Solution. Area $= \displaystyle\int_{-1}^{3} y^2 \, dy = \left(\frac{x^3}{3} \right)\Big|_{-1}^{3} = \frac{27}{3} - \frac{-1}{3} = \frac{28}{3}.$

Example (8.3-3). Let $F(x) = \displaystyle\int_0^x f(x) \, dx$, where the graph of f is given in the figure below.

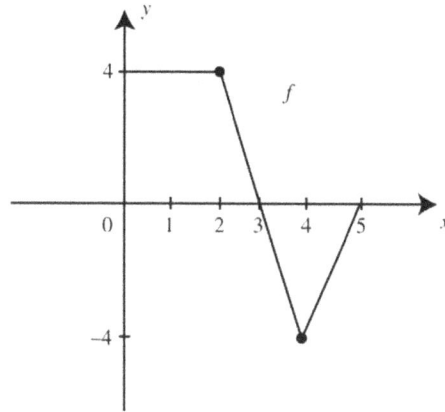

(1) Evaluate $F(0)$, $F(3)$ and $F(5)$.

(2) On what intervals is F increasing?

(3) At what value of t does F have the maximum value?

(4) On what intervals is F concave up?

Solution.

(1) By the Fundamental Theorem of Calculus,

$$F(0) = \int_0^0 f(x)\ dx = 0$$

$$F(3) = \int_0^3 f(x)\ dx = \frac{1}{2}(3+2)(4) = 10$$

$$F(5) = \int_0^3 f(x)\ dx + \int_3^5 f(x)\ dx = 10 + (-4) = 6$$

(2) Since $\displaystyle\int_3^5 f(x)\ dx < 0$, $f(x)$ is decreasing on $[3,5]$

(3) At $t = 3$, F has a maximum value.

(4) Since $F''(x) = f'(x)$ and $f(x)$ is increasing on $(4,5)$, thus $F(x)$ is concave up on $(4,5)$.

Example (8.4-1). Find the area of the regions bounded by the graphs of $f(x) = (x-1)^3$ and $g(x) = x - 1$.

Solution. Let $f(x) = g(x)$, we can find the point of intersection:

$$(x-1)^3 = x - 1 \ \Rightarrow\ (x-1)^3 - (x-1) = 0 \ \Rightarrow\ (x-1)[(x-1)^2 - 1] = 0$$

Thus $(x-1)x(x-2) = 0 \ \Rightarrow\ x = 0,\ 1,\ 2$. Then the area is

$$A = \int_0^1 \left[(x-1)^3 - (x-1)\right]\ dx + \int_0^1 \left[(x-1) - (x-1)^3\right]\ dx$$

$$= \left[\frac{(x-1)^4}{4} - \frac{(x-1)^2}{2}\right]_0^1 + \left[\frac{(x-1)^2}{2} - \frac{(x-1)^4}{4}\right]_1^2$$

$$= \frac{1}{4} + \frac{1}{4} = \frac{1}{2}$$

Alternatively, we can use $A = 2\int_0^1 \left[(x-1)^3 - (x-1)\right]\,\mathrm{d}x$ since the two regions are symmetric.

Example (8.4-2). Find the area of the regions bounded by the graphs of $x = y^2 - 4y$ and $x = y$.

Solution. The curves are shown in the figure below:

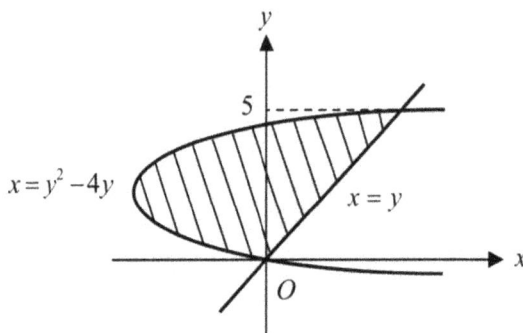

The point of intersection are given by

$$y^2 - 4y = y \implies y(y-5) = 0 \implies y = 0,\ 5$$

Then the area is given by

$$\begin{aligned}
A &= \int_0^5 \left[y - (y^2 - 4y)\right]\,\mathrm{d}y \\
&= \left(5y - y^2\right) \\
&= \left.\frac{5}{2}y^2 - \frac{1}{3}y^3\right|_0^5 \qquad\qquad = \left(\frac{125}{2} + \frac{125}{3}\right) - 0 = \frac{125}{6}
\end{aligned}$$

Example (8.5-1). The volume of the solid whose base is the region of the circle $x^2 + y^2 = 4$ whose cross sections taken perpendicular to the x-axis are squares.

Solution. The area of the square is $A(x) = (2y)^2 = 16 - 4x^2$

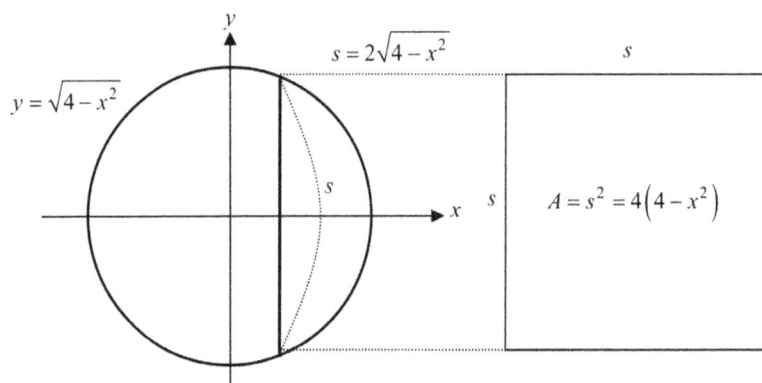

The volume is given by

$$V = \int_{-2}^{2} A(x)\, dx = \int_{-2}^{2} 16 - 4x^2 \, dx$$

$$= 16x - \frac{4}{3}x^3 \bigg|_{-2}^{2} = \left(32 - \frac{32}{3}\right) - \left(-32 + \frac{32}{3}\right)$$

$$= \frac{128}{3}$$

Example (8.5-2). Let R be the region in the first quadrant enclosed by $y = \sin x$, $y = \cos x$, and $x = 0$. Find the volume of the solid generated whose base is the region R and whose cross sections, perpendicular to the x-axis, are squares.

Solution. The point of intersection is given by $\sin x = \cos x$

Thus, $\tan x = 1 \;\Rightarrow\; x = \dfrac{\pi}{4}$

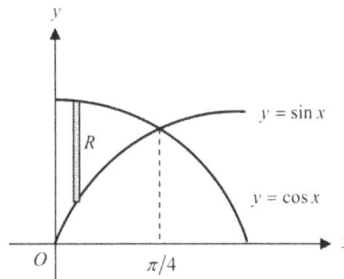

The volume is given by

$$V = \int_{0}^{\pi/4} (\cos x - \sin x)^2 \, dx = \int_{0}^{\pi/4} (1 - 2\sin x \cos x)\, dx = \int_{0}^{\pi/4} (1 - \sin 2x)\, dx$$

$$= x + \frac{1}{2}\cos 2x \bigg|_{0}^{\pi/4} = \frac{\pi}{4} - \frac{1}{2} = \frac{\pi - 2}{4}.$$

Example (8.5-3). The base of a solid is the region enclosed by the ellipse $\dfrac{x^2}{4} + \dfrac{y^2}{25} = 1$. The cross sections are perpendicular to the x-axis and are isosceles right triangles whose hypotenuses are on the ellipse. Find the volume of the solid.

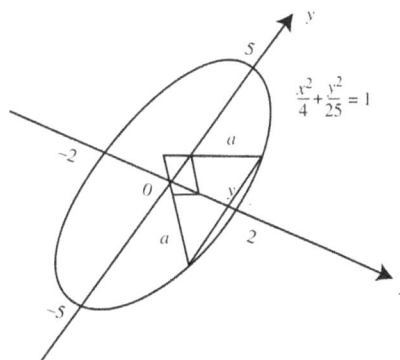

Solution. In the triangle $\sqrt{2}a = 2y \;\Rightarrow\; a = \sqrt{2}y$ Thus, $A(x) = \dfrac{1}{2}a^2 = y^2$

Since $\dfrac{x^2}{4} + \dfrac{y^2}{25} = 1$, then $y^2 = A(x) = 25 - \dfrac{25}{4}x^2$.

The volume is given by

$$V = \int_{-2}^{2} A(x)\, dx = \int_{-2}^{2} 25 - \frac{25}{4}x^2\, dx = 25x - \frac{25}{12}x^3 \Big|_{-2}^{2}$$

$$= \left(50 - \frac{200}{12}\right) - \left(-50 + \frac{200}{12}\right) = \frac{100}{3} - \left(-\frac{100}{3}\right) = \frac{200}{3}$$

Example (8.5-4). The base of a solid is the region enclosed by a triangle whose vertices are $(0,0)$, $(4,0)$, and $(0,2)$. The cross sections are semicircles perpendicular to the x-axis. Using a calculator, find the volume of the solid.

Solution. The equation passing $(0,2)$ and $(4,0)$ is $y = -\dfrac{1}{2}x + 2$

Then $A(x) = \dfrac{1}{2}\pi r = \dfrac{\pi}{2}\left(\dfrac{y}{2}\right)^2 = \dfrac{\pi}{2}\left(-\dfrac{1}{4}x + 1\right)^2$

The volume is given by

$$V = \int_{0}^{4} A(x)\, dx = \int_{0}^{4} \frac{\pi}{2}\left(-\frac{1}{4}x + 1\right)^2 dx = 2.094$$

Example (8.6-1). Let R be the region enclosed by $y = \tan x$, the x-axis, and $x = \dfrac{\pi}{3}$. Find the volume of the solid formed by revolving the region bounded by the graphs about the x-axis.

Solution. The volume is given by

$$V = \pi \int_{0}^{\pi/3} \tan^2 x\, dx = \pi \int_{0}^{\pi/3} \sec^2 x - 1\, dx$$

$$= \pi\, [\tan x - x]_{0}^{\pi/3} = \pi\left(\sqrt{3} - \frac{\pi}{3}\right)$$

Example (8.6-2). Find the volume of the solid generated by revolving about the y-axis the region in the first quadrant bounded by the graph of $y = x^2$, the y-axis, and the line $y = 6$.

Solution. $y = x^2 \Rightarrow x = \pm\sqrt{y}$ The volume is given by

$$V = \pi \int_{0}^{6} (\sqrt{y})^2\, dx = \pi \int_{0}^{6} y\, dx$$

$$= \pi \left[\frac{y^2}{2}\right]_{0}^{6} = 18\pi$$

Example (8.6-3). Find the volume of the solid generated when the region under the curve $y = x^2$ over the interval $[0, 2]$ is rotated about the line $x = 2$.

Solution. $y = x^2 \Rightarrow x = \pm\sqrt{y}$. The volume is given by

$$V = \pi \int_0^4 (x-2)^2 \, \mathrm{d}y = \pi \int_0^2 \left(\sqrt{y} - 2\right)^2 \, \mathrm{d}y$$

$$= \pi \int_0^4 \left(y - 4y^{\frac{1}{2}} + 4\right) \, \mathrm{d}y$$

$$= \pi \left[\frac{1}{2}y^2 - \frac{8}{3}y^{\frac{3}{2}} + 4y\right]_0^4$$

$$= \pi \left(8 - \frac{64}{3} + 16\right) - 0 = \frac{8}{3}\pi$$

Example (8.6-4). Using a calculator, find the volume of the solid generated by revolving about the line $y = -3$, the region bounded by the graph of $y = e^x$, the y-axis, and the lines $x = \ln 2$ and $y = -3$.

Solution. The volume is given by

$$V = \pi \int_0^{\ln 2} (y+3)^2 \, \mathrm{d}x$$

$$= \pi \int_0^{\ln 2} (e^x + 3)^2 \, \mathrm{d}x = 43.160$$

Example (8.7-1). Let R be the region in the first quadrant enclosed by the graph of $y = \sqrt{6x+4}$, the line $y = 2x$, and the y-axis. Find the volume of the solid generated when R is revolved about the x-axis.

Solution. The point of intersections is given by: $\sqrt{6x+4} = 2x$, then:

$$4x^2 = 6x + 4 \Rightarrow 2x^2 - 3x + 2 = (2x+1)(x-2) = 0 \Rightarrow x = -\frac{1}{2}, \ 2$$

Since R lies in the first quadrant, we only need to consider $x = 2$, then the point of intersection is $(2, 4)$.

Thus, the volume is given by

$$V = \pi \int_0^2 \left(\sqrt{6x+4}\right)^2 - (2x)^2 \, \mathrm{d}x = \pi \int_0^2 6x + 4 - 4x^2 \, \mathrm{d}x$$

$$= \pi \left(3x^2 + 4x - \frac{4}{3}x^3\right)\Big|_0^2 = \pi \left[\left(12 + 8 - \frac{32}{3}\right) - 0\right] = \frac{28\pi}{3}$$

Example (8.7-2). Using the Washer Method, find the volume of the solid generated by revolving the region bounded by $y = x^2$ and $x = y^2$ about the y-axis.

Solution. The point of intersections is given by: $x = y^2 = x^4$, then:

$$x^4 - x = 0 \Rightarrow x(x^3 - 1) = 0 \Rightarrow x = 0, \ 1$$

The point of intersection is $(0, 0)$ and $(1, 1)$.

The two curves are $x = y^2$ and $x = \sqrt{y}$. Thus, the volume is given by

$$V = \pi \int_0^1 (\sqrt{y})^2 - (y^2)^2 \; dy = \pi \int_0^1 y - y^4 \; dy$$

$$= \pi \left(\frac{y^2}{2} - \frac{y^5}{5} \right)\Big|_0^1 = \pi \left[\left(\frac{1}{2} - \frac{1}{5} \right) - 0 \right] = \frac{3\pi}{10}$$

Example (8.8-1). Find the arc length of the function $y = \dfrac{1}{3} \left(x^2 + 2 \right)^{\frac{3}{2}}$ over the interval $[0, 3]$.

Solution. Since $\dfrac{dy}{dx} = \dfrac{1}{2} \left(x^2 + 2 \right)^{\frac{1}{2}} \cdot (2x) = x\sqrt{x^2 + 2}$. The arc length is given by

$$L = \int_0^3 \sqrt{1 + \left(\frac{dy}{dx} \right)^2} \; dx = \int_0^3 \sqrt{1 + x^4 + 2x^2} \; dx = \int_0^3 \sqrt{(x^2 + 1)^2} \; dx$$

$$= \int_0^3 x^2 + 1 \; dx = \frac{x^3}{3} + x \Big|_0^3 = (9 + 3) - 0 = 12$$

Example (8.8-2). Find the arc length of the function $x = \dfrac{1}{3}\sqrt{y}(y - 3)$ from $y = 1$ to $y = 9$.

Solution. Since $x = \dfrac{1}{3} y^{\frac{3}{2}} - y^{\frac{1}{2}}$. Then $\dfrac{dx}{dy} = \dfrac{1}{2} y^{\frac{1}{2}} - \dfrac{1}{2} y^{-\frac{1}{2}} = \dfrac{1}{2} \left(\sqrt{y} - \dfrac{1}{\sqrt{y}} \right)$.

The arc length is given by

$$L = \int_1^9 \sqrt{1 + \left(\frac{dy}{dx} \right)^2} \; dy = \int_1^9 \sqrt{1 + \frac{1}{4}\left(\sqrt{y} - \frac{1}{\sqrt{y}} \right)^2} \; dy = \int_1^9 \frac{\sqrt{y}}{2} + \frac{1}{2\sqrt{y}} \; dy$$

$$= \int_1^9 \frac{1}{2} y^{\frac{1}{2}} + \frac{1}{2} y^{-\frac{1}{2}} \; dy = \frac{1}{3} y^{\frac{3}{2}} + y^{\frac{1}{2}} \Big|_1^9 = \frac{32}{3}$$

11.9   Unit 9 Solutions

Example (9.1-1). For parametric equations $x = t^2$ and $y = t^3 - 3t$ ($t \geq 0$)

   (a) Find the point where horizontal tangent occurs.

   (b) Find the point where vertical tangent occurs.

Solution.

   (a) $\dfrac{dy}{dt} = 3t^2 - 3 = 0 \;\Rightarrow\; t^2 = 1 \;\Rightarrow\; t = 1$, and $\dfrac{dx}{dt} = 2t$, at $t = 1$, $\dfrac{dx}{dt} \neq 0$

      Therefore, $x = 1$ and $y = -2 \;\Rightarrow\; (1, -2)$

   (b) $\dfrac{dx}{dt} = 2t = 0 \;\Rightarrow\; t = 0$ and $\dfrac{dy}{dt} = 3t^2 + 3 \neq 0$.

      Therefore, $x = 0$ and $y = 0 \;\Rightarrow\; (0,0)$

Example (9.1-2). If parametric equations $x = \sec t$ and $y = \tan t$ on $\left[-\dfrac{\pi}{2}, \dfrac{\pi}{2}\right]$, find the equation of tangent line at $t = \dfrac{\pi}{4}$.

Solution. At $t = \dfrac{\pi}{4}$, $x = \sec \dfrac{\pi}{4}] = \sqrt{2}$ and $\tan \dfrac{\pi}{4} = 1$, the point is $(\sqrt{2}, 1)$

The slope is $\left.\dfrac{dy}{dx}\right|_{t=\pi/4} = \left.\dfrac{dy/dt}{dx/dt}\right|_{t=\pi/4} = \left.\dfrac{\sec^2 t}{\sec t \tan t}\right|_{t=\pi/4} = \sec t|_{t=\pi/4} = \sqrt{2}$

Hence the equation of tangent is:

$$y - 1 = \sqrt{2}\left(x - \sqrt{2}\right) \;\Rightarrow\; y = \sqrt{2}x - 1.$$

Example (9.1-3). Find $\dfrac{dy}{dx}$ and $\dfrac{d^2y}{dx^2}$ at point $(1, 1)$ on the curve given by the parametric equations $x = t^2$, $y = t^3$.

Solution.

The first derivative is given by

$$y' = \frac{dy}{dx} = \frac{dy/dt}{dx/dt} = \frac{3t^2}{2t} = \frac{3}{2}t$$

For point $(1,1)$, we have $t = 1$, thus $\left.\dfrac{dy}{dx}\right|_{t=1} = \dfrac{3}{2}$

The second derivative is given by

$$\frac{d^2y}{dx^2} = \frac{dy'}{dx} = \frac{dy'/dt}{dx/dt} = \frac{3/2}{2t} = \frac{3}{4t}$$

When $t = 1$, $\left.\dfrac{d^2y}{dx^2}\right|_{t=1} = \dfrac{3}{4}$

Example (9.1-4). If $x = \ln(2t)$, $y = \ln(3t)^4$, find $\dfrac{d^2y}{dx^2}$ in terms of t

Solution.

Since $y = 4\ln(3t)$, $\dfrac{dy}{dt} = 4 \cdot \dfrac{3}{3t} = \dfrac{4}{t}$, and $\dfrac{dx}{dt} = \dfrac{2}{2t} = \dfrac{1}{t}$

The first derivative is given by

$$y' = \frac{dy}{dx} = \frac{dy/dt}{dx/dt} = 4$$

Thus $\dfrac{d^2y}{dx^2} = 0$

Example (9.2-1). Find arc length of the function over the indicated interval.

(a) $x = \dfrac{1}{3}t^3$, $y = \dfrac{1}{2}t^2$, $0 \le t \le 1$

(b) $x = e^{-t}\cos t$, $y = e^{-t}\sin t$, $0 \le t \le \dfrac{\pi}{2}$

Solution.

(a) $\dfrac{dx}{dt} = t^2$, $\dfrac{dy}{dt} = t$, then:

$$L = \int_0^1 \sqrt{t^4 + t^2}\; dt = \int_0^1 t\sqrt{t^2 + 1}\; dt$$

Let $u = t^2 + 1$, then $du = 2t\; dt \;\Rightarrow\; \dfrac{1}{2}\,du = t\; dt$

When $t = 0$, $u = 1$, and when $t = 1$, $u = 2$, thus the integration can be rewritten as

$$L = \int_0^1 t\sqrt{t^2+1}\; dt = \frac{1}{2}\int_1^2 \sqrt{u}\; du = \frac{1}{3}\left(u^{\frac{3}{2}}\Big|_1^2\right) = \frac{2\sqrt{2}-1}{3}$$

(b) The derivative of x and y with respect to t is

$$\begin{cases} \dfrac{dx}{dt} &= -e^{-t}\cos t - e^{-t}\sin t = e^{-t}(\cos t - \sin t) \\ \dfrac{dy}{dt} &= -e^{-t}\sin t + e^{-t}\cos t = e^{-t}(\sin t + \cos t) \end{cases}$$

Thus $\left(\dfrac{dx}{dt}\right)^2 + \left(\dfrac{dy}{dt}\right)^2$ can be expressed as:

$$\left(\frac{dx}{dt}\right)^2 + \left(\frac{dy}{dt}\right)^2 = e^{-2t}\left[(\cos t - \sin t)^2 + (\sin t + \cos t)^2\right] = 2e^{-2t}$$

Thus the arc length is:

$$L = \int_0^{\pi/2} \sqrt{2e^{-2t}}\; dt = \int_0^{\pi/2} \sqrt{2}\,e^{-t}\; dt = -\sqrt{2}\,e^{-t}\Big|_0^{\frac{\pi}{2}} = -\sqrt{2}\,e^{\frac{\pi}{2}} + \sqrt{2}$$

Example (9.3-1). A particle moves in an elliptical path so that its position at any time $t \ge 0$ is given by $4\sin t, 2\cos t$.

(a) Find the velocity and acceleration vectors.

(b) Find the velocity, acceleration, speed, and direction of motion at $t = \dfrac{\pi}{4}$.

Solution.

(a) Velocity $\mathbf{v}(t) = \left\langle \dfrac{d}{dt}(4\sin t), \dfrac{d}{dt}(2\cos t) \right\rangle = \langle 4\cos t, -2\sin t \rangle$

Acceleration $\mathbf{a}(t) = \left\langle \dfrac{d}{dt}(4\cos t), \dfrac{d}{dt}(-2\sin t) \right\rangle = \langle -4\sin t, -2\cos t \rangle$

(b) Velocity $= \mathbf{v}(\dfrac{\pi}{4}) = \left\langle 4\cos\dfrac{\pi}{4}, -2\sin\dfrac{\pi}{4} \right\rangle = \left\langle 2\sqrt{2}, -\sqrt{2} \right\rangle$

Acceleration $= \mathbf{v}(\dfrac{\pi}{4}) = \left\langle -2\sqrt{2}, -\sqrt{2} \right\rangle$

Speed $= \left| \mathbf{v}(\dfrac{\pi}{4}) \right| = \sqrt{\left(2\sqrt{2} \right)^2 + \left(-\sqrt{2} \right)^2} = \sqrt{10}$

Example (9.3-2). A particle moves in the plane with velocity vector $\mathbf{v}(t) = \langle t - 3\pi\cos(\pi t), 2t - \pi\sin(\pi t) \rangle$. At $t = 0$, the particle is at the point $(1, 5)$. Use your calculator, find:

(a) The position of the particle at $t = 4$.

(b) The total distance traveled by the particle from $t = 0$ to $t = 4$?

Solution.

(a) Displacement $= \left\langle \displaystyle\int_0^4 t - 3\pi\cos(\pi t)\,dt, \int_0^4 2t - \pi\sin(\pi t)\,dt \right\rangle = \langle 8, 16 \rangle$

at $t = 0$, the particle is at $(1, 5)$, thus at $t = 4$, the particle is at $(1 + 8, 5 + 16) = (9, 21)$

(b) Distance $= \displaystyle\int_0^4 \sqrt{(t - 3\pi\cos(\pi t))^2 + (2t - \pi\sin(\pi t))^2}\,dt \approx 33.533.$

Example (9.3-3). For time $t \geq 0$, a particle moves in the xy-plane with position $(x(t), y(t))$ and velocity vector $\left\langle (t-1)e^{t^2}, \sin t^{1.25} \right\rangle$. At time $t = 0$, the position of the particle is $(-2, 5)$. Use your calculator, find:

(a) The speed of the particle at time $t = 1.2$ and the acceleration vector of the particle at time $t = 1.2$.

(b) The total distance traveled by the particle over the time interval $0 \leq t \leq 1.2$.

(c) Find the coordinates of the point at which the particle is farthest to the left for $t \geq 0$. Explain why there is no point at which the particle is farthest to the right for $t \geq 0$.

Solution.

(a) Speed $= \sqrt{(x'(1.2))^2 + (y'(1.2))^2} = 1.271$

At time $t = 1.2$, the speed of the particle is 1.271.

Acceleration $\mathbf{a}(1.2) = \langle x''(1.2), y''(1.2) \rangle \langle 6.247, 0.405 \rangle$

At time $t = 1.2$, the acceleration vector of the particle is $\langle 6.247, 0.405 \rangle$.

(b) Distance $= \int_0^{1.2} \sqrt{\left((t-1)e^{t^2}\right)^2 + (\sin t^{1.25})^2}\ dt \approx 1.010$.

The total distance traveled by the particle over the time interval $0 \le t \le 1.2$ is 1.010.

(c) $x'(t) = (t-1)e^{t^2} = 0 \Rightarrow t = 1$

Because $x'(t) < 0$ for $0 \le t < 1$ and $x'(t) > 0$ for $t > 1$, the particle is farthest to the left at time $t = 1$, then

$$x(t) = -2 + \int_0^1 (t-1)e^{t^2}\ dt = -2.604$$

$$y(t) = 5 + \int_0^1 \sin t^{1.25}\ dt = 5.410$$

Thus, the coordinate farthest to the left is $(-2.604, 5.410)$

Since $x'(t) > 0$ when $t > 1$, the particle is moving to the right when $x > 1$.

In addition, $x(2) = -2 + \int_0^2 (t-1)e^{t^2}\ dt > -2 = x(0)$, so the particle's motion extends to the right of its initial position after time $t = 1$.

Therefore, there is no point at which the particle is farthest to the right.

Example (9.4-1). Find $\dfrac{dy}{dx}$ and slope of tangent at $\theta = \dfrac{\pi}{6}$ for $r = 1 + \cos\theta$

Solution. We can see $f'(\theta) = -\sin\theta$ and substitute in the equation.

$$\frac{dy}{dx} = \frac{f'(\theta)\sin\theta + f(\theta)\cos\theta}{f'(\theta)\cos\theta - f(\theta)\sin\theta} = \frac{-\sin^2\theta + (1+\cos\theta)\cos\theta}{-\sin\theta\cos\theta - (1+\cos\theta)\sin\theta}$$

$$= \frac{-\sin^2\theta + \cos\theta + \cos^2\theta}{-2\sin\theta\cos\theta-} = -\left(\frac{\cos 2\theta + \cos\theta}{\sin 2\theta + \sin\theta}\right)$$

By the derivative, we may find that at $\theta = \dfrac{\pi}{6}$, $\dfrac{dy}{dx} = -\dfrac{\frac{1}{2} + \frac{\sqrt{3}}{2}}{\frac{\sqrt{3}}{2} + \frac{1}{2}} = -1$

Example (9.4-2). The polar curve r is given by $r(\theta) = 1 - \sin\theta$ for $0 \le \theta \le 2\pi$. Find the area in the second quadrant enclosed by the coordinate axes and the graph of r.

Solution. The area is given by

$$A = \frac{1}{2}\int_{\pi/2}^{\pi} (1 - \sin\theta)^2\ d\theta = \frac{1}{2}\int_{\pi/2}^{\pi} (1 - 2\sin\theta + \sin^2\theta)\ d\theta$$

$$= \frac{1}{2}\int_{\pi/2}^{\pi}\left(1 - 2\sin\theta + \frac{1-\cos 2\theta}{2}\right)\ d\theta$$

$$= \frac{1}{2}\int_{\pi/2}^{\pi}\left(\frac{3}{2} - 2\sin\theta - \frac{1}{2}\cos 2\theta\right)\ d\theta = \frac{1}{2}\left(\frac{3}{2}\theta + 2\cos\theta - \frac{1}{4}\sin 2\theta\right)\Big|_{\frac{\pi}{2}}^{\pi}$$

$$= \frac{1}{2}\left[\left(\frac{3}{2}\pi + 2 - 0\right) - \left(\frac{3}{4}\pi + 0 - 0\right)\right] = \frac{1}{2}\left(\frac{3}{4}\pi + 2\right) = \frac{3}{8}\pi + 1$$

Example (9.4-3). Find the area of the region in the plane enclosed by the cardioid $r = 2(1 + \cos\theta)$.

Solution. The area is given by

$$
\begin{aligned}
A &= \frac{1}{2}\int_0^{2\pi} 4(1 + \cos\theta)^2 \, d\theta = 2\int_0^{2\pi}(1 - 2\cos\theta + \cos^2\theta)\, d\theta \\
&= \int_0^{2\pi}(2 + 4\cos\theta + 1 + \cos 2\theta)\, d\theta \\
&= \int_0^{2\pi}(3 + 4\cos\theta + \cos 2\theta)\, d\theta = \left(3\theta + 4\sin\theta + \frac{1}{2}\sin 2\theta\right)\Big|_0^{2\pi} \\
&= [(6\pi + 0 - 0) - 0] = 6\pi
\end{aligned}
$$

Example (9.4-4). Find the area inside the smaller loop of the limacon $r = 2\cos\theta + 1$.

Solution. $r = 2\cos\theta + 1 \;\Rightarrow\; \cos\theta = -1,\; \theta = \dfrac{2\pi}{3}$ or $\dfrac{4\pi}{3}$. Then:

$$
A = \frac{1}{2}\int_{2\pi/3}^{4\pi/3}(2\cos\theta + 1)^2 \, d\theta \approx 0.544
$$

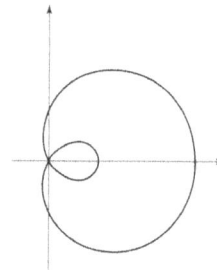

Example (9.4-5). Let R be the region inside the graph of the polar curve $r = 2$ and outside the graph of the polar curve $r = 2(1 - \sin\theta)$.

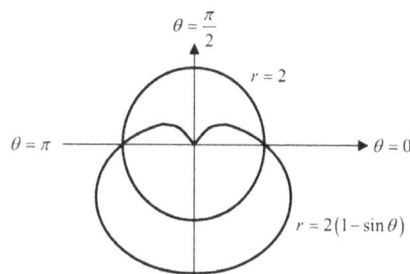

Solution.

The area of the semicircle is $\dfrac{1}{2}\pi(2)^2 = 2\pi$

The area of the shaded region $= \pi - \displaystyle\int_0^{\pi}\frac{1}{2}\cdot 4(1 - \sin\theta)^2 \, d\theta = \pi - \int_0^{\pi} 2(1 - \sin\theta)^2 \, d\theta$

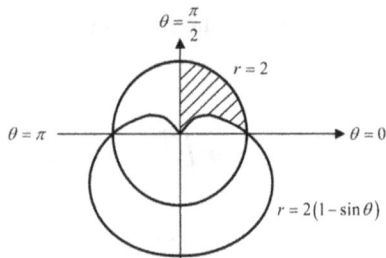

$$\frac{A}{2} = \pi - 2 \int_0^{\pi/2} (1 - \sin\theta)^2 \, d\theta = \pi - 2 \int_0^{\pi/2} (1 - 2\sin\theta + \sin^2\theta) \, d\theta$$

$$= \pi - 2 \int_0^{\pi/2} \left(1 - 2\sin\theta + \frac{1 - \cos 2\theta}{2}\right) d\theta$$

$$= \pi - 2 \int_0^{\pi/2} \left(\frac{3}{2} - 2\sin\theta - \frac{1}{2}\cos 2\theta\right) d\theta = \pi - 2 \left(\frac{3}{2}\theta + 2\cos\theta - \frac{1}{4}\sin 2\theta\right)\Big|_0^{\pi/2}$$

$$= \pi - 2 \left[\left(\frac{3}{4}\pi + 0 - 0\right) - (0 + 2 - 0)\right] = \pi - 2 \left(\frac{3}{4}\pi - 2\right) = 4 - \frac{\pi}{2}$$

Now we obtain the area $A = 8 - \pi$.

Example (9.4-6).

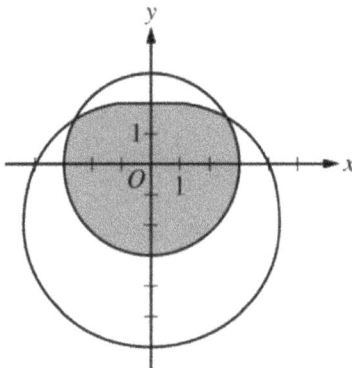

The graphs of the polar curves $r = 3$ and $r = 4 - 2\sin\theta$ are shown in the figure above. The curves intersect when $\theta = \dfrac{\pi}{6}$ and $\theta = \dfrac{5\pi}{6}$.

(a) Let S be the shaded region that is inside the graph of $r = 3$ and also inside the graph of $r = 4 - 2\sin\theta$. Find the area of S.

(b) A particle moves along the polar curve $r = 4 - 2\sin\theta$ so that at time t seconds, $\theta = t^2$. Find the time t in the interval $1 \leq t \leq 2$ for which the x-coordinate of the particle's position is -1.

(c) For the particle described in part (b), find the position vector in terms of t. Find the velocity vector at time $t = 1.5$.

Solution.

(a) Area $= 6\pi + \dfrac{1}{2} \displaystyle\int_{\pi/6}^{5\pi/6} (4 - 2\sin\theta)^2 \, d\theta = 24.709$

(b) $x = r\cos\theta \implies x(\theta) = (4 - 2\sin\theta)\cos\theta$

$x(t) = \left(4 - 2\sin\left(t^2\right)\right)\cos\left(t^2\right)$

when $x(t) = -1$, $t = 1.428$

(c) $y(\theta) = r \sin \theta = (4 - 2 \sin \theta) \sin \theta$

$y(t) = (4 - 2 \sin(t^2)) \sin(t^2)$

Position vector $= \langle x(t), y(t) \rangle = \langle (4 - 2 \sin(t^2)) \cos(t^2), (4 - 2 \sin(t^2)) \sin(t^2) \rangle$

$\mathbf{v}(1.5) = \langle x'(1.5), y'(1.5) \rangle = \langle -8.072, -1.673 \rangle$

11.10   Unit 10 Solutions

Example (10.1-1). Write the formula for the following sequences and determine their limits (if exist).

$$\frac{1}{3}, \frac{2}{4}, \frac{3}{5}, \frac{4}{6}, \frac{5}{7}, \cdots$$

Solution. Here, $a_n = \dfrac{n}{n+2}$. Then the limit is:

$$\lim_{n\to\infty} \frac{n}{n+2} = \lim_{n\to\infty} \frac{n+2-2}{n+2} = \lim_{n\to\infty}\left(1 - \frac{2}{n+2}\right) = \lim_{n\to\infty} 1 - \lim_{n\to\infty}\frac{2}{n+2} = 1 - 0 = 1.$$

Thus, the sequence converges to 1.

Example (10.1-2). Write the formula for the following sequences and determine their limits (if exist).

$$1, -\frac{2}{2}, \frac{3}{4}, -\frac{4}{8}, \frac{5}{16}, \cdots$$

Solution. The n-th term of the sequence is: $a_n = \dfrac{(-1)^{n-1}n}{2^{n-1}}$. Then the limit is:

Since $-n \le (-1)^{n-1}n \le n$, we can write $-\dfrac{n}{2^{n-1}} \le \dfrac{(-1)^{n-1}n}{2^{n-1}} \le \dfrac{n}{2^{n-1}}$

Using **L'Hospital's Rule**, we obtain:

$$\lim_{n\to\infty}\frac{\pm n}{2^{n-1}} = \pm\lim_{n\to\infty}\frac{n}{2^{n-1}\ln 2} = 0.$$

Hence, by the squeezing theorem, the limit of the initial sequence is $\displaystyle\lim_{n\to\infty}\frac{(-1)^{n-1}n}{2^{n-1}} = 0$

Example (10.1-3). Does the sequence $\left\{\dfrac{2n+3}{5n-7}\right\}$ converge or diverge?

Solution. Divide by the highest power in the numerator and denominator:

$$\lim_{n\to\infty}\frac{2n+3}{5n-7} = \lim_{n\to\infty}\frac{\dfrac{2n+3}{n}}{\dfrac{5n-7}{n}} = \lim_{n\to\infty}\frac{2 + \dfrac{3}{n}}{5 - \dfrac{7}{n}} = \frac{2}{5}.$$

Since the limit is finite, the given sequence converges to $\dfrac{2}{5}$.

You can also use the L'Hopital's rule to find the limit.

Example (10.1-4). Determine whether the sequence $\left\{\sqrt{n+2} - \sqrt{n+1}\right\}$ converges or diverges.

Solution. First, we *rationalize the denominator*, by multiplying $\sqrt{n+2}+\sqrt{n+1}$ for both sides:

$$\lim_{n\to\infty}\left(\sqrt{n+2}-\sqrt{n+1}\right) = \lim_{n\to\infty}\left(\sqrt{n+2}-\sqrt{n+1}\right)\cdot\frac{\sqrt{n+2}+\sqrt{n+1}}{\sqrt{n+2}+\sqrt{n+1}}$$

$$= \lim_{n\to\infty}\frac{\left(\sqrt{n+2}\right)^2-\left(\sqrt{n+1}\right)^2}{\sqrt{n+2}+\sqrt{n+1}}$$

$$= \lim_{n\to\infty}\frac{n+2-(n+1)}{\sqrt{n+2}+\sqrt{n+1}}$$

$$= \lim_{n\to\infty}\frac{1}{\sqrt{n+2}+\sqrt{n+1}}=0$$

This means that the sequence converges.

Example (10.1-5). Does the sequence $\left\{\dfrac{n^2}{2^n}\right\}$ converge or diverge?

Solution. As L'Hospital's Rule yields:

$$\lim_{x\to\infty}\left(\pm\frac{x}{2^{x-1}}\right) = \pm\lim_{x\to\infty}\frac{x}{2^{x-1}} = \pm\lim_{x\to\infty}\frac{1}{2^{x-1}\ln 2}=0.$$

Since the limit is finite, the given sequence converges.

Example (10.1-6). Determine whether the series $\displaystyle\sum_{n=1}^{\infty} n$ converges or diverges.

Solution.

The partial sum is: $S_n = 1+2+3+\cdots+n = \dfrac{1}{2}n(n+1)$

By taking the limit: $\displaystyle\lim_{n\to\infty} S_n = \infty$, thus the series diverges.

Example (10.1-7). Determine whether the series $\displaystyle\sum_{n=1}^{\infty}(-1)^{n+1}\cdot 2 = 2-2+2\cdots$ converges or diverges.

Solution.

The partial sum is: $S_n = \begin{cases} 2, & n \text{ is odd} \\ 0, & n \text{ is even} \end{cases}$

By taking the limit: $\displaystyle\lim_{n\to\infty} S_n$ does not exist, thus the series diverges.

Example (10.2-1). Determine whether the following series converges or diverges and find the value that the series converge to:

(a) $\displaystyle\sum_{n=1}^{\infty}\frac{1}{n(n+1)}.$

(b) $\displaystyle\sum_{n=4}^{\infty}\frac{1}{n^2-4n+3}.$

Solution.

(a) The partial sum is given by:

$$S_n = \sum_{k=1}^{n} \frac{1}{k+1} = \sum_{k=1}^{n} \left(\frac{1}{k} - \frac{1}{k+1} \right)$$

$$= \left(1 - \frac{1}{2} \right) + \left(\frac{1}{2} - \frac{1}{3} \right) + \left(\frac{1}{3} - \frac{1}{4} \right) + \ldots + \left(\frac{1}{n} - \frac{1}{n+1} \right)$$

$$= 1 - \frac{1}{n+1}$$

The infinite sum is $\lim_{n \to \infty} S_n = \lim_{n \to \infty} \left(1 - \frac{1}{n+1} \right) = 1 - 0 = 1$. Thus, the series converges.

(b) The partial sum is given by:

$$S_n = \sum_{k=4}^{n} \frac{1}{k^2 - 4k + 3} = \sum_{k=4}^{n} \frac{1}{(k-3)(k-1)} = \sum_{k=4}^{n} \frac{1}{2} \left(\frac{1}{k-3} - \frac{1}{k-1} \right)$$

$$= \frac{1}{2} \left[\left(1 - \frac{1}{3} \right) + \left(\frac{1}{2} - \frac{1}{4} \right) + \left(\frac{1}{3} - \frac{1}{5} \right) + \left(\frac{1}{4} - \frac{1}{6} \right) + \ldots + \left(\frac{1}{n-4} - \frac{1}{n-2} \right) + \left(\frac{1}{n-3} - \frac{1}{n-1} \right) \right]$$

$$= \frac{1}{2} \left(1 + \frac{1}{2} - \frac{1}{n-2} - \frac{1}{n-1} \right)$$

The infinite sum is:

$$\lim_{n \to \infty} S_n = \lim_{n \to \infty} \frac{1}{2} \left(1 + \frac{1}{2} - \frac{1}{n-2} - \frac{1}{n-1} \right)$$

$$= \frac{1}{2} \left(1 + \frac{1}{2} - 0 \right) = \frac{3}{4}.$$

Thus, the series converges.

Example (10.2-2). For an arbitrary geometric sequence $\{a_n\}$ with first term a, and common ratio r ($r \neq 0$, 1), find the partial sum S_n for the series.

Solution.

The n-th partial sum can be written as:

$$S_n = a + ar + ar^2 + ar^3 + \ldots + ar^n \qquad (1)$$

By multiplying a to both sides, we get:

$$rS_n = ar + ar^2 + ar^3 + \ldots + ar^n + ar^{n+1} \qquad (2)$$

Subtracting equation (2) from (1):

$$(1 - r) S_n = a - ar^n = a(1 - r^n)$$

Since the common ratio is generally not 1, thus we may solve that:

$$S_n = \frac{a(1 - r^n)}{1 - r}$$

Example (10.2-3). Find the sum of the series $\displaystyle\sum_{n=1}^{\infty} a_n = 1 - \frac{1}{\sqrt{2}} + \frac{1}{2} - \frac{1}{2\sqrt{2}} + \frac{1}{4} - \frac{1}{4\sqrt{2}} + \dots$

Solution.

This is a geometric series with common ratio: $r = -\dfrac{1}{\sqrt{2}}$

Since $|r| < 1$, then the series converges to:

$$\frac{a}{1-r} = \frac{1}{1 - \left(-\dfrac{1}{\sqrt{2}}\right)} = \frac{1}{1 + \dfrac{1}{\sqrt{2}}} = \frac{\sqrt{2}}{\sqrt{2} + 1} = \frac{\sqrt{2}\left(\sqrt{2} - 1\right)}{\left(\sqrt{2} + 1\right)\left(\sqrt{2} - 1\right)} = 2 - \sqrt{2}$$

Example (10.2-4). Consider a recurring decimal $0.1313131313\cdots$

(a) State the decimal using a geometric series.

(b) Find the exact fractional value of the decimal using convergence.

Solution.

(a) The decimal can be written as:

$$0.13 + 0.13 \times \frac{1}{100} + 0.13 \times \frac{1}{100^2} + 0.13 \times \frac{1}{100^3} + \dots$$

This is a geometric series with first term 0.13, and common ratio $r = \dfrac{1}{100}$

Since $|r| < 1$, then the series converges.

(b) By the formula of geometric series, the series converges to:

$$\frac{a}{1-r} = \frac{0.13}{1 - \frac{1}{100}} = \frac{0.13}{0.99} = \frac{13}{99}$$

Example (10.2-5). Find the range of x such that $\displaystyle\sum_{n=0}^{\infty} x^n$ converges, and evaluate the value of the series.

Solution.

$\displaystyle\sum_{n=0}^{\infty} x^n = 1 + x + x^2 + \dots$ is a geometric series with common ratio x.

If the series converge, then $|x| < 1$.

By the formula of geometric series, the series converges to:

$$\frac{1}{1-x} = (1-x)^{-1}$$

Example (10.2-6). Show that the harmonic series diverges.

Solution.

The harmonic series can be written as: $\displaystyle\sum_{n=1}^{\infty} \frac{1}{n} = 1 + \frac{1}{2} + \frac{1}{3} + \frac{1}{4} + \dots$

The terms can also be grouped into the following style:

$$\sum_{n=1}^{\infty} \frac{1}{n} = 1 + \frac{1}{2} + \left(\frac{1}{3} + \frac{1}{4}\right) + \left(\frac{1}{5} + \frac{1}{6} + \frac{1}{7} + \frac{1}{8}\right) + \dots$$

$$> 1 + \frac{1}{2} + \left(\frac{1}{4} + \frac{1}{4}\right) + \left(\frac{1}{8} + \frac{1}{8} + \frac{1}{8} + \frac{1}{8}\right) + \dots$$

$$= 1 + \frac{1}{2} + \frac{1}{2} + \frac{1}{2} + \dots$$

$$= \lim_{k \to \infty} \left(1 + \frac{k}{2}\right)$$

Since the limit approaches infinity, thus we may deduce that the harmonic series diverges.

** *The **integral test** can also be used to show the divergence of harmonic series, which will be discussed in later.*

Example (10.3-1). Verify whether the series diverge or not.

(a) $\displaystyle\sum_{n=1}^{\infty} \frac{n}{2n+3}$
(b) $\displaystyle\sum_{n=1}^{\infty} \frac{e^n}{n^2}$
(c) $\displaystyle\sum_{n=1}^{\infty} \frac{1}{\cos\left(\frac{1}{n}\right)}$

Solution.

(a) $\displaystyle\lim_{n\to\infty} \frac{n}{2n+3} = \frac{1}{2} \neq 0$, the series diverge.

(b) Using **L'Hospital's Rule**, $\displaystyle\lim_{n\to\infty} \frac{e^n}{n^2} = \lim_{n\to\infty} \frac{e^n}{2n} = \lim_{n\to\infty} \frac{e^n}{2} = \infty$, the series diverge.

(c) $\displaystyle\lim_{n\to\infty} \frac{1}{\cos\left(\frac{1}{n}\right)} = \lim_{n\to\infty} \frac{1}{\cos 0} = 1 \neq 0$, the series diverge.

Example (10.3-2). Verify whether the series diverge or not.

(a) $\displaystyle\sum_{n=1}^{\infty} \frac{1}{n^e}$
(b) $\displaystyle\sum_{n=1}^{\infty} \frac{2}{n\sqrt{n}}$
(c) $\displaystyle\sum_{n=1}^{\infty} n^{\frac{1}{3}}$
(d) $\displaystyle\sum_{n=1}^{\infty} \frac{5}{\sqrt{n}}$

Solution.

(a) This is a p-series with $p = e$, since $p > 1$, the series converge.

(b) $\displaystyle\sum_{n=1}^{\infty} \frac{2}{n\sqrt{n}} = 2\sum_{n=1}^{\infty} \frac{1}{n^{\frac{3}{2}}}$ This is a p-series with $p = \frac{3}{2}$, since $p > 1$, the series converge.

(c) $\displaystyle\sum_{n=1}^{\infty} n^{\frac{1}{3}} = \sum_{n=1}^{\infty} \frac{1}{n^{-\frac{1}{3}}}$ This is a p-series with $p = -\frac{1}{3}$, since $p \leq 1$, the series diverge.

(d) $\displaystyle\sum_{n=1}^{\infty} \frac{5}{\sqrt{n}} = 5\sum_{n=1}^{\infty} \frac{1}{n^{\frac{1}{2}}}$ This is a p-series with $p = \frac{1}{2}$, since $p \leq 1$, the series diverge.

Example (10.3-3). Determine whether the series $\displaystyle\sum_{n=1}^{\infty} \frac{n^2-1}{n^4}$ converges or diverges.

Solution.

Note that the series is similar to $\displaystyle\sum_{n=1}^{\infty} \frac{n^2}{n^4} = \sum_{n=1}^{\infty} \frac{1}{n^2}$.

We use the comparison test. Note that for all positive integer n:

$$\frac{n^2 - 1}{n^4} < \frac{n^2}{n^4} = \frac{1}{n^2}$$

As $\displaystyle\sum_{n=1}^{\infty} \frac{1}{n^2}$ is a p-series with $p = 2 > 1$, it converges.

Hence, the given series also converges by the comparison test.

Example (10.3-4). Determine whether the series $\displaystyle\sum_{n=1}^{\infty} \frac{n^2}{n^3 - 3}$ converges or diverges.

Solution.

Note that the series is similar to $\displaystyle\sum_{n=1}^{\infty} \frac{n^2}{n^3} = \sum_{n=1}^{\infty} \frac{1}{n}$.

We may find that $n^3 - 3 < n^3$ for all positive integer n. Then, by the comparison test:

$$\frac{1}{n^3 - 3} > \frac{1}{n^3} \implies \frac{n^2}{n^3 - 3} > \frac{n^2}{n^3} = \frac{1}{n}.$$

Since $\displaystyle\sum_{n=1}^{\infty} \frac{1}{n}$ is a harmonic series, which diverges.

Hence, the given series also diverges by the comparison test.

Example (10.3-5). Determine whether the series $\displaystyle\sum_{n=1}^{\infty} \frac{e^{\frac{1}{n}}}{n^2}$ converges or diverges.

Solution.

Note that the series is similar to $\displaystyle\sum_{n=1}^{\infty} \frac{e}{n^2} = e \sum_{n=1}^{\infty} \frac{1}{n^2}$.

We may find that $e^{\frac{1}{n}} < e$ for all positive integer n. Then, by the comparison test:

$$\sum_{n=1}^{\infty} \frac{e^{\frac{1}{n}}}{n^2} \leq \sum_{n=1}^{\infty} \frac{e}{n^2} = e \sum_{n=1}^{\infty} \frac{1}{n^2}.$$

As $\displaystyle e \sum_{n=1}^{\infty} \frac{1}{n^2}$ is a p-series with $p = 2 > 1$, it converges.

Hence, the given series also converges by the comparison test.

Example (10.3-6). Determine whether the series $\displaystyle\sum_{n=1}^{\infty} \frac{2 + \cos n}{n^3}$ converges or diverges.

Solution.

Note that the series is similar to $\displaystyle\sum_{n=1}^{\infty} \frac{2}{n^3}$.

We may find that $\cos n \leq 1 \Rightarrow 2 + \cos n \leq 3$ for all positive integer n, then:

$$\frac{2 + \cos n}{n^3} \leq \frac{3}{n^3}$$

As $\displaystyle 3 \sum_{n=1}^{\infty} \frac{1}{n^3}$ is a p-series with $p = 3 > 1$, it converges.

Hence, the given series also converges by the comparison test.

Example (10.3-7). Determine whether the series $\displaystyle\sum_{n=0}^{\infty} \frac{2^n \sin^2 (5n)}{4^n + \cos^2 (3n)}$ converges or diverges.

Solution.

Note that the series is similar to $\displaystyle\sum_{n=0}^{\infty} \frac{2^n}{4^n} = \sum_{n=0}^{\infty} \left(\frac{1}{2}\right)^n$.

We may find that $\sin^2 (5n) \leq 1$, $\cos^2 (3n) \geq 0 \Rightarrow 2^n \sin^2 (5n) \leq 2^n$, $4^n + \cos^2 (3n) \geq 4^n$ for all n, then:

$$\frac{2^n \sin^2 (5n)}{4^n + \cos^2 (3n)} \leq \frac{2^n}{4^n} = \left(\frac{1}{2}\right)^n$$

As $\displaystyle\sum_{n=0}^{\infty} \left(\frac{1}{2}\right)^n$ is a geometric series with $|r| = \frac{1}{2} < 1$, it converges.

Hence, the given series also converges by the comparison test.

Example (10.3-8). Determine whether the series $\displaystyle\sum_{n=1}^{\infty} \frac{n^2+1}{n^4}$ converges or diverges.

Solution.

Note that the series is similar to $\displaystyle\sum_{n=1}^{\infty} \frac{n^2}{n^4} = \sum_{n=1}^{\infty} \frac{1}{n^2}$.

If we use the comparison test. We may find that $\dfrac{n^2+1}{n^4} > \dfrac{n^2}{n^4} = \dfrac{1}{n^2}$ for all positive integer n:

However, $\displaystyle\sum_{n=1}^{\infty} \frac{1}{n^2}$ is a p-series with $p = 2 > 1$, it converges. The test is *inconclusive*.

Therefore, we need to use the **limit comparison test**:

$$L = \lim_{n\to\infty} \frac{\dfrac{n^2+1}{n^4}}{\dfrac{n^2}{n^4}} = \lim_{n\to\infty} \frac{n^2+1}{n} = 1$$

Since the limit is finite, and given that $\displaystyle\sum_{n=1}^{\infty} \frac{1}{n^2}$ converges, thus $\displaystyle\sum_{n=1}^{\infty} \frac{n^2+1}{n^4}$ also converges.

Example (10.3-9). Determine whether the series $\displaystyle\sum_{n=1}^{\infty} \frac{3n-1}{2n^3-4n+5}$ converges or diverges.

Solution.

Note that the series is similar to $\displaystyle\sum_{n=1}^{\infty} \frac{3n}{2n^3} = \frac{3}{2}\sum_{n=1}^{\infty} \frac{1}{n^2}$. By the limit test, we get:

$$L = \lim_{n\to\infty} \frac{\frac{3n-1}{2n^3-4n+5}}{\frac{3n}{2n^3}} = \lim_{n\to\infty} \frac{(3n-1)\,2n^3}{3n\,(2n^3-4n+5)} = \lim_{n\to\infty} \frac{6n^4-2n^3}{6n^4-12n^2+15n} = 1.$$

Since $\dfrac{3}{2}\displaystyle\sum_{n=1}^{\infty} \frac{1}{n^2}$ is a p-series with $p = 2 > 1$, it converges.

Hence, the given series also converges by the test.

Example (10.3-10). Determine whether the series $\displaystyle\sum_{n=1}^{\infty} \frac{\sqrt{n}}{2n^2 + n + 5}$ converges or diverges.

Solution.

Note that the series is similar to $\displaystyle\sum_{n=1}^{\infty} \frac{\sqrt{n}}{2n^2} = \frac{1}{2}\sum_{n=1}^{\infty} \frac{1}{n^{\frac{3}{2}}}$. By the limit test, we get:

$$L = \lim_{n \to \infty} \frac{\frac{\sqrt{n}}{n+n+}}{\frac{\sqrt{n}}{2n^2}} = \lim_{n \to \infty} \frac{2n^2}{2n^2 + n + 5} = 1.$$

Since $\dfrac{1}{2}\displaystyle\sum_{n=1}^{\infty} \dfrac{1}{n^{\frac{3}{2}}}$ is a p-series with $p = \dfrac{3}{2} > 1$, it converges.

Hence, the given series also converges by the test.

Example (10.3-11). Determine whether the series $\displaystyle\sum_{n=1}^{\infty} \sin\left(\frac{1}{n}\right)$ converges or diverges.

(*Hint: Consider the limit:* $\displaystyle\lim_{x \to 0} \frac{\sin x}{x} = 1$ *and use the substitution* $x = \dfrac{1}{n}$)

Solution.

Using the limit and the substitution $x = \dfrac{1}{n}$ we obtain

$$\lim_{x \to 0} \frac{\sin x}{x} = \lim_{n \to \infty} \frac{\sin\left(\frac{1}{n}\right)}{\frac{1}{n}} = 1$$

Then we can use the limit comparison test with the harmonic series $\displaystyle\sum_{n=1}^{\infty} \frac{1}{n}$

Since the harmonic series diverges, and the limit $L = 1$.

Hence, the given series also diverges by the test.

Example (10.3-12). Determine whether the series $\displaystyle\sum_{n=1}^{\infty} \frac{3^n}{n^2}$ converges or diverges.

Solution. By the ratio test, we have:

$$\rho = \lim_{n \to \infty} \frac{a_{n+1}}{a_n} = \lim_{n \to \infty} \frac{\frac{3^{n+1}}{(n+1)^2}}{\frac{3^n}{n^2}}$$

$$= \lim_{n \to \infty} \left[\frac{3^{n+1}}{3^n} \cdot \frac{n^2}{(n+1)^2}\right]$$

$$= \lim_{n \to \infty} \left[3\left(\frac{n}{n+1}\right)^2\right] = 3\lim_{n \to \infty} \left(\frac{n+1-1}{n+1}\right)^2$$

$$= 3\lim_{n \to \infty} \left(1 - \frac{1}{n+1}\right)^2 = 3.$$

Since $\rho > 1$, the series diverges.

Example (10.3-13). Determine whether the series $\displaystyle\sum_{n=1}^{\infty} \frac{n^3}{(\ln 3)^n}$ converges or diverges.

Solution. By the ratio test, we have:

$$\rho = \lim_{n\to\infty} \frac{a_{n+1}}{a_n} = \lim_{n\to\infty} \frac{\dfrac{(n+1)^3}{(\ln 3)^{n+1}}}{\dfrac{n^3}{(\ln 3)^n}}$$

$$= \lim_{n\to\infty} \left[\frac{(\ln 3)^n}{(\ln 3)^{n+1}} \cdot \frac{(n+1)^3}{n^3} \right]$$

$$= \frac{1}{\ln 3} \lim_{n\to\infty} \left(\frac{n+1}{n} \right)^3 = \frac{1}{\ln 3}.$$

Since $\rho > 1$, the series diverges.

Example (10.3-14). Determine whether the series $\dfrac{(1!)^2}{2!} + \dfrac{(2!)^2}{4!} + \cdots + \dfrac{(n!)^2}{(2n)!} + \cdots$ converges or diverges.

Solution. By the ratio test, we have:

$$\rho = \lim_{n\to\infty} \frac{a_{n+1}}{a_n} = \lim_{n\to\infty} \frac{\dfrac{((n+1)!)^2}{(2(n+1))!}}{\dfrac{(n!)^2}{(2n)!}} = \lim_{n\to\infty} \left[\frac{((n+1)!)^2}{(n!)^2} \cdot \frac{(2n)!}{(2n+2)!} \right]$$

$$= \lim_{n\to\infty} \left[\frac{\cancel{(n!)^2}(n+1)^2}{\cancel{(n!)^2}} \cdot \frac{\cancel{(2n)!}}{\cancel{(2n)!}(2n+1)(2n+2)} \right]$$

$$= \lim_{n\to\infty} \frac{(n+1)^2}{(2n+1)(2n+2)} = \lim_{n\to\infty} \frac{n^2+2n+1}{4n^2+2n+4n+2} = \frac{1}{4}.$$

Since $\rho < 1$, the series converges.

Example (10.3-15). Determine whether the series $\displaystyle\sum_{n=1}^{\infty} \frac{n^n}{2^{3n-1}}$ converges or diverges.

Solution. By the root test, we have:

$$\lim_{n\to\infty} \sqrt[n]{\frac{n^n}{2^{3n-1}}} = \lim_{n\to\infty} \frac{n}{2^{3-\frac{1}{n}}} = \lim_{n\to\infty} \frac{n}{2^3 \cdot 2^{-\frac{1}{n}}} = \lim_{n\to\infty} \frac{n \cdot 2^{\frac{1}{n}}}{8} = \infty > 1.$$

Since the root is greater than 1, the series diverges.

Example (10.3-16). Determine whether the series $\dfrac{1}{(\ln 2)^2} + \dfrac{1}{(\ln 3)^3} + \dfrac{1}{(\ln 4)^4} + \cdots$ converges or diverges.

Solution. The series is: $\displaystyle\sum_{n=2}^{\infty} \frac{1}{(\ln n)^n}$. By the root test, we have:

$$\lim_{n\to\infty} \sqrt[n]{a_n} = \lim_{n\to\infty} \sqrt[n]{\frac{1}{(\ln n)^n}} = \lim_{n\to\infty} \frac{1}{\ln n} = 0 < 1.$$

Since the root is less than 1, the series converges.

Example (10.3-17). Determine whether the series $\displaystyle\sum_{n=0}^{\infty} ne^{-n}$ converges or diverges.

Solution. Using the integral test, the improper integral is given by

$$\int_0^\infty xe^{-x}\,\mathrm{d}x = \lim_{b\to\infty} \int_0^b xe^{-x}\,\mathrm{d}x.$$

Using **integration by parts**, we get

$$
\begin{aligned}
\int_0^\infty xe^{-x}\,\mathrm{d}x &= \lim_{b\to\infty} \int_0^b xe^{-x}\mathrm{d}x \\
&= \lim_{b\to\infty} \left[(-xe^{-x})\big|_0^b + \int_0^b e^{-x}\mathrm{d}x \right] \\
&= \lim_{b\to\infty} (-xe^{-x} - e^{-x})\big|_0^b \\
&= (-0 - 0) - (-0 - 1) = 1
\end{aligned}
$$

The integral converges; hence the series converges.

Example (10.3-18). Determine whether the series $\displaystyle\sum_{n=2}^{\infty} \frac{1}{n\ln n}$ converges or diverges.

Solution. Using the integral test, the improper integral is given by

$$\int_2^\infty \frac{1}{x\ln x}\,\mathrm{d}x = \lim_{b\to\infty} \int_2^b \frac{1}{x\ln x}\,\mathrm{d}x.$$

Let $u = \ln x \Rightarrow \mathrm{d}u = \frac{1}{x}\,\mathrm{d}x$, then:

$$
\begin{aligned}
\lim_{b\to\infty} \int_2^b \frac{1}{x\ln x}\,\mathrm{d}x &= \lim_{b\to\infty} \int_{\ln 2}^b \frac{1}{u}\,\mathrm{d}u \\
&= \lim_{b\to\infty} (\ln|u|)\big|_{\ln 2}^b \\
&= \ln(\ln|b|) - \ln(\ln|2|) = \infty
\end{aligned}
$$

The integral is divergent and so the series is also divergent.

Example (10.3-19). Use the integral test to show that the p-series $\sum\limits_{n=1}^{\infty} \dfrac{1}{n^p}$ converges for $p > 1$ and diverges for $p \leq 1$.

Solution. We consider the corresponding integral $\displaystyle\int_1^{\infty} \dfrac{1}{x^p}\,\mathrm{d}x = \int_1^{\infty} x^{-p}\,\mathrm{d}x$

(1) For $p > 1 \Rightarrow p - 1 > 0$, the integral is given by

$$\int_1^{\infty} x^{-p}\,\mathrm{d}x = \lim_{b \to \infty} \int_1^b x^{-p}\,\mathrm{d}x = \lim_{b \to \infty} \left(\frac{x^{-p+1}}{-p+1}\right)\bigg|_1^b$$

$$= \frac{1}{1-p} \cdot \lim_{b \to \infty} \left(\frac{1}{x^{p-1}}\right)\bigg|_1^b$$

$$= \frac{1}{1-p} \cdot \lim_{b \to \infty} \left(\frac{1}{b^{p-1}} - 1\right)$$

$$= \frac{1}{1-p} \cdot (0 - 1)$$

$$= \frac{1}{p-1}$$

The improper integral converges; thus, the series converge for $p > 1$.

(2) For $p < 1 \Rightarrow 1 - p > 0$, the integral is given by

$$\int_1^{\infty} x^{-p}\,\mathrm{d}x = \lim_{b \to \infty} \int_1^b x^{-p}\,\mathrm{d}x = \lim_{b \to \infty} \left(\frac{x^{-p+1}}{-p+1}\right)\bigg|_1^b$$

$$= \frac{1}{1-p} \cdot \lim_{b \to \infty} \left(\frac{1}{x^{p-1}}\right)\bigg|_1^b$$

$$= \frac{1}{1-p} \cdot \lim_{b \to \infty} \left(\frac{1}{b^{p-1}} - 1\right)$$

$$= \frac{1}{1-p} \cdot \lim_{b \to \infty} \left(b^{1-p} - 1\right)$$

$$= \infty$$

The improper integral diverges; thus, the series converge for $p < 1$.

(3) For $p = 1$

$$\int_1^{\infty} x^{-1}\,\mathrm{d}x = \lim_{b \to \infty} \int_1^b \frac{1}{x}\,\mathrm{d}x$$

$$= \lim_{b \to \infty} (\ln x)\big|_1^b$$

$$= \lim_{b \to \infty} (\ln b - \ln 1)$$

$$= \lim_{b \to \infty} (\ln b)$$

$$= \infty$$

The improper integral diverges; thus, the series converge for $p = 1$.

Example (10.4-1). Determine whether the series $\sum_{n=1}^{\infty} (-1)^n \frac{2}{3n}$ converges or diverges.

Solution. By the alternating series test, let $a_n = \frac{2}{3n}$, then:

$$(1)\ a_{n+1} = \frac{2}{3n+3} < a_n \qquad (2)\ \lim_{n\to\infty} a_n = \lim_{n\to\infty} \frac{2}{3n} = 0.$$

Thus the series converges.

Example (10.4-2). Determine whether the series $\sum_{n=1}^{\infty} (-1)^n \frac{2}{3n}$ is absolutely convergent, conditionally convergent or divergent.

Solution. From **Example 1**, we have already checked that the alternating series converges.

Taking the absolute value, we find that $\sum_{n=1}^{\infty} |a_n| = \frac{2}{3} \sum_{n=1}^{\infty} \frac{1}{n}$, which is a harmonic series.

Since harmonic series diverge. Thus, the series is conditionally convergent.

Example (10.4-3). Determine whether the series $\sum_{n=1}^{\infty} \frac{(-1)^{n+1}}{n!}$ is absolutely convergent, conditionally convergent or divergent.

Solution.

We can use the ratio test. First we take $|a_n| = \left| \frac{(-1)^{n+1}}{n!} \right| = \frac{1}{n!}$, and the ration ρ is

$$\rho = \lim_{n\to\infty} \left| \frac{a_{n+1}}{a_n} \right| = \lim_{n\to\infty} \frac{\dfrac{1}{(n+1)!}}{\dfrac{1}{n!}} = \lim_{n\to\infty} \frac{n!}{(n+1)!} = \lim_{n\to\infty} \frac{1}{n+1} = 0.$$

Since $\sum_{n=1}^{\infty} |a_n|$ is convergent, then $\sum_{n=1}^{\infty} \frac{(-1)^{n+1}}{n!}$ is absolutely convergent.

Example (10.4-4). Determine whether the series $\sum_{n=1}^{\infty} \frac{(-1)^{n+1}}{5n-1}$ is absolutely convergent, conditionally convergent or divergent.

Solution.

By the alternating series test, we find that (1) $a_{n+1} < a_n$, and (2) $\lim_{n\to\infty} \frac{1}{5n-1} = 0$.

Thus, the alternating series converges. Then, taking $|a_n|$, we may use the comparison test:

Given that $\frac{1}{5n-1} > \frac{1}{5n}$,

Since $\sum_{n=1}^{\infty} \frac{1}{5n} = \frac{1}{5} \sum_{n=1}^{\infty} \frac{1}{n}$ diverges (harmonic series).

We may deduce that the series $\sum_{n=1}^{\infty} \dfrac{1}{5n-1}$ diverges.

and $\sum_{n=1}^{\infty} \dfrac{(-1)^{n+1}}{5n-1}$ is conditionally convergent.

Example (10.5-1). For a power series $\sum_{n=0}^{\infty} x^n = 1 + x + x^2 + x^3 + \cdots$, find the range of value of x such that the series converge.

Solution.

$1 + x + x^2 + x^3 + \cdots$ is a geometric series with common ratio x, which converges when $|x| < 1$.

Example (10.5-2). Show that $\sum_{n=0}^{\infty} \dfrac{(x+3)^n}{n!}$ is convergent for all value of x.

Solution. By the ratio test:

$$\rho = \lim_{n\to\infty} \left| \frac{a_{n+1}}{a_n} \right| = \lim_{n\to\infty} \left| \frac{\frac{(x+3)^{n+1}}{(n+1)!}}{\frac{(x+3)^n}{n!}} \right| = \lim_{n\to\infty} \frac{x+3}{n+1} = (x+3) \cdot \lim_{n\to\infty} \frac{1}{n+1} = 0$$

The series $\sum_{n=0}^{\infty} x^n = 1 + x + x^2 + x^3 + \cdots$ is a geometric series with common ration x.

The ratio is always 0, ($\rho < 1$, independent with x), then the series is convergent for all x.

Example (10.5-3). Determine the radius of convergence for the power series $f(x) = \sum_{n=1}^{\infty} a_n (x-c)^n$.

Solution. By the ratio test:

$$\rho = \lim_{n\to\infty} \left| \frac{a_{n+1}(x-c)^{n+1}}{a_n(x-c)^n} \right| = \lim_{n\to\infty} \left| \frac{a_{n+1}}{a_n}(x-c) \right| = |x-c| \cdot \lim_{n\to\infty} \left| \frac{a_{n+1}}{a_n} \right|$$

The series is convergent when $\rho = |x-c| \cdot \lim_{n\to\infty} \left| \dfrac{a_{n+1}}{a_n} \right| < 1 \implies |x-c| < \lim_{n\to\infty} \left| \dfrac{a_n}{a_{n+1}} \right|$

Then the radius of convergence is given by:

$$R = \lim_{n\to\infty} \left| \frac{a_n}{a_{n+1}} \right|.$$

Note that the end point should be checked separately.

Example (10.5-4). Determine the interval of convergence for the power series $\sum_{n=0}^{\infty} n x^n$.

Solution.

The center of the interval is located at $x = 0$. By the **Radius Corollary**:

$$R = \lim_{n \to \infty} \left| \frac{a_n}{a_{n+1}} \right| = \lim_{n \to \infty} \frac{n}{n+1} = \lim_{n \to \infty} \frac{1}{1 + \frac{1}{n}} = 1.$$

For the endpoints:

(1) If $x = -1$, the series $\sum_{n=0}^{\infty} (-1)^n n$ diverges.

(2) If $x = 1$, the series $\sum_{n=0}^{\infty} n$ also diverges.

Therefore, the interval of convergence for the series is $-1 < x < 1$.

Example (10.5-5). For what values of x does the series $\sum_{n=0}^{\infty} \frac{x^n}{n+1}$ converge?

Solution.

The center of the interval is located at $x = 0$. By the **Radius Corollary**:

$$R = \lim_{n \to \infty} \left| \frac{a_n}{a_{n+1}} \right| = \lim_{n \to \infty} \frac{\frac{1}{n+1}}{\frac{1}{n+2}} = \lim_{n \to \infty} \frac{n+2}{n+1} = 1.$$

For the endpoints:

(1) If $x = -1$, the series $\sum_{n=0}^{\infty} \frac{(-1)^n}{n+1}$ converges (alternate series test).

(2) If $x = 1$, the series $\sum_{n=0}^{\infty} \frac{1}{n+1}$ diverges (compare with harmonic series).

Therefore, the interval of convergence for the series is $-1 \leq x < 1$.

Example (10.5-6). Determine the radius and interval of convergence for the power series $\sum_{n=0}^{\infty} \frac{(-1)^n (x-2)^n}{n^2}$.

Solution.

The center of the interval is located at $x = 2$. By the **Radius Corollary**:

$$R = \lim_{n \to \infty} \left| \frac{a_n}{a_{n+1}} \right| = \lim_{n \to \infty} \frac{\frac{1}{n^2}}{\frac{1}{(n+1)^2}} = \lim_{n \to \infty} \frac{(n+1)^2}{n^2} = \lim_{n \to \infty} \left(\frac{n+1}{n} \right)^2 = 1.$$

For the endpoints:

(1) If $x = 2 - 1 = 1$, the series $\sum_{n=0}^{\infty} \frac{(-1)^n}{n^2}$ converges (alternate series test).

(2) If $x = 2 + 1 = 3$, the series $\sum_{n=0}^{\infty} \frac{1}{n^2}$ converges (p-series).

Therefore, the interval of convergence for the series is $1 \leq x \leq 3$.

Example (10.6-1). For the Maclaurin Series of a function:

$$(x) = a_0 + a_1 x + a_2 x^2 + a_3 x^3 + \ldots a_n x^n + \ldots = P(x)$$

Show that for any positive integer n, $a_n = \dfrac{f^{(n)}(0)}{n!}$

Solution.

By $P^{(n)}(x) = f^{(n)}(x)$, we may find that:

For $n = 0$, $P(0) = a_0 = f(0)$;   For $n = 1$, $P'(0) = a_1 = f'(0)$

For $n = 2$, $P''(0) = 2 \times 1 \cdot a_2 = f''(0) \implies a_2 = \dfrac{f''(0)}{2!}$

For $n = 3$, $P'''(0) = 3 \times 2 \times 1 \cdot a_3 = f'''(0) \implies a_3 = \dfrac{f'''(0)}{3!}$

Then for any n, we have:

$$P^{(n)}(0) = n! \cdot a_n = f^{(n)}(0) \implies a_n = \dfrac{f^{(n)}(0)}{n!}$$

Example (10.6-2). Find the Maclaurin Series for $f(x) = e^x$, and show that the series converges for any real value of x.

Solution.

For any positive integer n, $f^{(n)}(0) = e^x|_{x=0} = 1$, then by the expansion formula:

$$f(x) = f(0) + f'(0) \cdot x + \frac{f''(0)}{2!} \cdot x^2 + \frac{f'''(0)}{3!} \cdot x^3 + \cdots + \frac{f^{(n)}(0)}{n!} \cdot x^n + \cdots$$

$$= 1 + x + \frac{x^2}{2!} + \frac{x^3}{3!} + \cdots + \frac{x^n}{n!} + \cdots = \sum_{n=0}^{\infty} \frac{x^n}{n!}$$

By the **ratio test**, we may find that the ratio *rho* is:

$$\rho = \lim_{n \to \infty} \left| \frac{a_{n+1}}{a_n} \right| = \lim_{n \to \infty} \frac{\dfrac{x^{n+1}}{(n+1)!}}{\dfrac{x^n}{n!}} = \lim_{n \to \infty} \frac{x}{n+1} = 0 < 1.$$

Since ρ is always less than 1, then the series is convergent for any real value of x.

Example (10.6-3). Find the Maclaurin Series for $f(x) = e^{x^2}$.

Solution.

Using the *substitution* $u = x^2$, then $f(x) = e^u$.

The Maclaurin Series for $f(x)$ is centered at $x = 0$, where $u = 0$ as well.

Then We can use the Maclaurin Series for e^u directly.

$$e^{x^2} = e^u = \sum_{n=0}^{\infty} \frac{u^n}{n!} = 1 + u + \frac{u^2}{2!} + \frac{u^3}{3!} + \cdots$$

$$= 1 + x^2 + \frac{x^4}{2!} + \frac{x^6}{3!} + \cdots = \sum_{n=0}^{\infty} \frac{x^{2n}}{n!}$$

*Note: Remember to **check the center of convergence** before substitution.*

Example (10.6-4). Find the first 3 non-zero terms of the Maclaurin Series for $f(x) = \sin\left(5x + \frac{\pi}{4}\right)$.

Solution.

Using the *substitution* $u = 5x + \frac{\pi}{4}$, then $f(x) = \sin u$.

The Maclaurin Series for $f(x)$ is centered at $x = 0$, where $u = \frac{\pi}{4} \neq 0$.

We **CANNOT** use the method of substation. Instead, we can use our **traditional way**.

$$f(0) = \sin\left(\frac{\pi}{4}\right) = \frac{\sqrt{2}}{2}, \quad f'(0) = 5\cos\left(\frac{\pi}{4}\right) = \frac{5\sqrt{2}}{2}, \quad f''(0) = -25\sin\left(\frac{\pi}{4}\right) = -\frac{25\sqrt{2}}{2}$$

By THEOREM 10.24, we may find that

$$P(x) = \frac{\sqrt{2}}{2} + \frac{5\sqrt{2}}{2}x - \frac{25\sqrt{2}}{2\cdot 2!}x^2 = \frac{\sqrt{2}}{2} + \frac{5\sqrt{2}}{2}x - \frac{25\sqrt{2}}{4}x^2$$

Example (10.6-5). Find the Maclaurin series for $\cos^2 x$ and write it in sigma notation.

Solution.

By the trigonometric identity $\cos^2 x = \frac{1}{2}(1 + \cos 2x)$

The Maclaurin Series for $\cos 2x$ can be written by substitution method:

$$\cos 2x = \sum_{n=0}^{\infty} \frac{(-1)^n (2x)^{2n}}{(2n)!} = \sum_{n=0}^{\infty} \frac{(-1)^n 2^{2n} x^{2n}}{(2n)!}.$$

Therefore, the Maclaurin series for $\cos^2 x$ is

$$1 + \cos 2x = 1 + \sum_{n=0}^{\infty} \frac{(-1)^n 2^{2n} x^{2n}}{(2n)!} = 2 + \sum_{n=1}^{\infty} \frac{(-1)^n 2^{2n} x^{2n}}{(2n)!}$$

$$\cos^2 x = \frac{1 + \cos 2x}{2} = 1 + \sum_{n=1}^{\infty} \frac{(-1)^n 2^{2n-1} x^{2n}}{(2n)!}.$$

Example (10.6-6). Determine the Taylor series for $f(x) = 3x^2 - 6x + 5$ about the point $x = 1$.

Solution.

Compute the derivatives:

$$f'(x) = 6x - 6, \quad f''(x) = 6, \quad f'''(x) = 0.$$

And $f^{(n)}(x) = 0$ for any $n \geq 4$, then for $x = 1$

$$f(1) = 2, \quad f'(1) = 0, \quad f''(1) = 6.$$

Hence, the Taylor expansion for the given function is:

$$f(x) = \sum_{n=0}^{\infty} \frac{f^{(n)}(1)\cdot(x-1)^n}{n!} = 2 + \frac{6(x-1)^2}{2!} = 2 + 3(x-1)^2.$$

Example (10.6-7). Use the 4th degree Maclaurin polynomial for $\ln(1+x)$ to approximate $\ln(1.1)$.

Solution.

The Maclaurin series for $\ln(1+x)$ is given by:

$$\ln(1+x) = x - \frac{x^2}{2} + \frac{x^3}{3} - \frac{x^4}{4} + \cdots$$

To approximate $\ln(1.1)$, we need to let $x = 0.1$, then the approximation is:

$$\ln(1+0.1) \approx 0.1 - \frac{0.1^2}{2} + \frac{0.1^3}{3} - \frac{0.1^4}{4} = 0.0953$$

** **Alternative Solution**: We may use the Maclaurin Series for $\ln x$ and let $x = 1.1$ to solve.

Example (10.6-8). The function f is defined by $f(x) = \sin(x^2)$. What are the first four nonzero terms of the Taylor series for f', about $x = 0$?

Solution.

By the **method of substitution**, we may write:

$$f(x) = x^2 - \frac{x^6}{3!} + \frac{x^{10}}{5!} - \frac{x^{14}}{7!} + \cdots$$

(*Note that we need to check the center before using substitution*)

Differentiating both sides, we get the first for terms as:

$$f'(x) \approx 2x - \frac{6x^5}{3!} + \frac{10x^9}{5!} - \frac{14x^{13}}{7!}$$

Example (10.6-9). The Maclaurin series for the function f is given by $\displaystyle\sum_{n=1}^{\infty} \frac{(-2)^n x^{2n}}{n}$. Find $f^{(4)}(0)$.

Solution.

Within the Maclaurin series, $f^{(4)}(0)$ appears in the term $\dfrac{f^{(4)}(0)}{4!}x^4$

In the power series for f, the term with 4th power appears when $2n = 4 \Rightarrow n = 2$.

For this value of n, the coefficient is:

$$\frac{(-2)^2}{2} = \frac{f^{(4)}(0)}{4!} = 2 \Longrightarrow f^{(4)}(0) = 2 \times 4! = 48$$

Example (10.6-10). Find the Maclaurin series for $(1+x)^n$.

Solution.

Compute the derivatives:

$$f'(x) = n(1+x)^{n-1}$$
$$f''(x) = n(n-1)(1+x)^{n-2}$$
$$f'''(x) = n(n-1)(n-2)(1+x)^{n-3}$$
$$\vdots$$
$$f^{(n)}(x) = n(n-1)(n-2)(n-3)...(n-k+1)(1+x)^{n-k}$$

For $x = 0 \Rightarrow 1 + x = 1$, we obtain:

$$f'(0) = n$$
$$f''(0) = n(n-1)$$
$$f'''(0) = n(n-1)(n-2)$$
$$\vdots$$
$$f^{(n)}(0) = n(n-1)(n-2)(n-3)...(n-k+1)$$

Hence, the series expansion can be written in the form:

$$(1+x)^n = 1 + nx + \frac{n(n-1)}{2!}x^2 + \frac{n(n-1)(n-2)}{3!}x^2 + \cdots$$
$$= \sum_{k=1}^{\infty} \frac{n(n-1)(n-2)...(n-k+1)}{k!}x^k$$

Which is known as the **binomial theorem.**

In particular, if n is a positive integer, then there must be a term with $(n+1)$-th degree as:

$$\frac{n(n-1)(n-2)... \times 1 \times 0}{(n+1)!}x^{n-(n+1)} = \frac{n! \times 0}{(n+1)!}x^{-1} = 0$$

For all succussing terms, the coefficient must be 0, then the series expansion is finite:

$$(1+x)^n = 1 + nx + \frac{n(n-1)}{2!}x^2 + \frac{n(n-1)(n-2)}{3!}x^2 + ... + \frac{n(n-1)(n-2)... \times 1}{n!}x^n$$
$$= \binom{n}{0} + \binom{n}{1}x + \binom{n}{2}x^2 + \binom{n}{3}x^3 + \cdots \binom{n}{n}x^n$$

Example (10.7-1). If the series $\sum_{n=1}^{\infty} \frac{(-1)^n}{5n+1}$ is approximated by the partial sum with 15 terms, what is the alternating series error bound?

Solution.

By using Error $\leq |a_{n+1}|$, we get:

$$\text{Error} \leq |a_{n+1}| = \left| \frac{1}{5(n+1)+1} \right| = \left| \frac{1}{5 \times 16 + 1} \right| = \frac{1}{81}$$

Example (10.7-2). The function f is defined by the power series $\displaystyle\sum_{n=1}^{\infty} \frac{(-1)^n\, x^n}{(2n+1)!}$. Show that the first three terms of the power series approximate $f(1)$ with error less than $\dfrac{1}{4000}$.

Solution.

By using Error $\leq |a_{n+1}|$, we get:

$$\text{Error} \leq |a_{n+1}| = \left| \frac{x^4}{(2 \times 4 - 1)!} \right| = \left| \frac{1}{7!} \right| = \frac{1}{5040} < \frac{1}{4000}$$

Example (10.7-3). Based on the alternating series error bound, what is the least value of m such that:

$$\left| \ln(1.1) - \sum_{n=1}^{m} \frac{(-1)^{n+1}}{n} \cdot (0.1)^n \right| < \frac{1}{5000}$$

Solution.

By using Error $\leq |a_{m+1}|$, we get:

$$\text{Error} \leq |a_{m+1}| = \left| \frac{1}{m+1} \cdot 0.1^{m+1} \right| = \frac{1}{(m+1) \cdot 10^{m+1}} < \frac{1}{5000}$$

When $m = 2$, $\dfrac{1}{(m+1) \cdot 10^{m+1}} = \dfrac{1}{3000} > \dfrac{1}{5000}$,

When $m = 3$, $\dfrac{1}{(m+1) \cdot 10^{m+1}} = \dfrac{1}{40000} < \dfrac{1}{5000}$.

Thus, the least value of m is 3.

Example (10.7-4). Based on the Lagrange error bound, what is the least value of m such that:

$$\left| \ln(1.1) - \sum_{n=1}^{m} \frac{(-1)^{n+1}}{n} \cdot (0.1)^n \right| < \frac{1}{5000}$$

Solution.

The Lagrange Error Bound is given by:

$$\text{Error} = R_4(x) \leq \max \left| f^{(4+1)}(x) \right| \cdot \frac{|0.3 - 0|^{4+1}}{(4+1)!}$$

$$= 1 \times \frac{0.3^5}{5!}$$

$$= 2.025 \times 10^{-5}$$

Example (10.7-5). Let f be a function that has derivatives of all orders for all real numbers and let $P_3(x)$ be the third-degree Taylor Polynomial for f about $x = 0$. $\left| f^{(n)}(x) \right| \leq \dfrac{n}{n+1}$ for all $1 \leq n \leq 5$ and for all values of x. What is the smallest value of k for which the Lagrange Error

Bound guarantees that $|f(1) - P_3(1)| \leq k$?

$$\left| \ln(1.1) - \sum_{n=1}^{m} \frac{(-1)^{n+1}}{n} \cdot (0.1)^n \right| < \frac{1}{5000}$$

Solution. The Lagrange Error Bound is given by:

$$R_3(x) \leq \max \left| f^{(4)}(x) \right| \cdot \frac{|1-0|^4}{4!} \leq \frac{4}{5} \times \frac{1}{24} = \frac{1}{30}$$

Thus, the smallest value of k is $k = \dfrac{1}{30}$.

Example (10.7-6). The function f has derivatives of all orders for all real numbers. Values of f and its first four derivatives at $x = 2$ are given in the table.

x	$f(x)$	$f'(x)$	$f''(x)$	$f'''(x)$	$f^{(4)}(x)$
2	6	-12	18	-24	34

The fourth derivative of f satisfies the inequality $\left| f^{(4)}(x) \right| \leq 48$ for all $x > 1$. Use the Lagrange error bound to show that the approximation for $f(1.5)$ using the third-degree Taylor Polynomial differs from $f(1.5)$ by no more than $\dfrac{1}{8}$.

Solution. The Lagrange Error Bound is given by:

$$R_3(x) \leq \max \left| f^{(4)}(x) \right| \cdot \frac{|1.5-2|^4}{4!} \leq 48 \times \frac{0.5^4}{24} = \frac{1}{8}$$

Thus, $|f(1.5) - P_3(1.5)| \leq \dfrac{1}{8}$